Workbook

Machining Fundamentals

Ninth Edition

Instructor's Annotated Workbook

by

John R. Walker

Publisher
The Goodheart-Willcox Company, Inc.
Tinley Park, IL
www.g-w.com

Manufactured in the United States of America.

ISBN 978-1-61960-214-4

1 2 3 4 5 6 7 8 9 – 14 – 18 17 16 15 14 13

Instructor's Annotated Workbook

ISBN 978-1-61960-216-8

1 2 3 4 5 6 7 8 9 – 14 – 18 17 16 15 14 13

Introduction

This workbook is designed for use with the text ***Machining Fundamentals***. It has been designed to help you develop a better understanding of machining technology and its many career opportunities. The chapters in the workbook correspond to those in the text and should be completed after reading the appropriate text chapter.

Each chapter of the workbook contains reviews of the textbook chapters to enhance your understanding of textbook content. The various types of questions include matching, multiple choice, fill-in-the-blank, identification, and short answer.

Reading ***Machining Fundamentals*** and using this workbook will help you acquire a working knowledge of the various machining technologies and their applications. Answering the questions for each chapter will help you master the technical knowledge presented in the text.

The world of machining is very complex. There is little or no need for unskilled workers. The more you know about machining, the better your chances will be to secure employment in this trade. As a machinist, you will be doing work you enjoy and will feel proud of the high quality products produced. At the same time, you will help make the world a better place to live.

Table of Contents

CHAPTER 1

An Introduction to Machining Technology

Name ______________________ Date __________ Class __________

Learning Objectives

After studying this chapter, you will be able to:

- Discuss how modern machine technology affects the workforce.
- Give a brief explanation of the evolution of machine tools.
- Provide an overview of machining processes.
- Explain how CNC machining equipment operates.
- Describe the role of the machinist.

Carefully study the chapter, then answer the following questions.

reciprocating 1. The back-and-forth motion of the bow lathe is known as _____ motion.

D 2. The first true machine tool was _____.

A. a form of the lathe called a boring mill
B. brought about with the need for perfectly bored cylinders
C. water-powered
D. All of the above.
E. None of the above.

B 3. Henry Maudslay designed and constructed the _____.

A. first true machine tool
B. granddaddy of all modern chip-making machine tools
C. boring mill
D. All of the above.
E. None of the above.

4. What great period in history could *not* have taken place without the advent of the steam engine?
Industrial Revolution

A 5. Eli Whitney _____.
A. devised a system for mass production in the 1820s
B. was an English inventor and manufacturer
C. invented the standard measuring system used in the United States
D. All of the above.
E. None of the above.

6. List seven types of power sources.
Any order: hand power, foot power, animal power, water power, steam power, central electrical power, individual electrical power.

A 7. A drill press removes metal through the use of _____.
A. spiral flutes pulling the tool into the work
B. a drill rotating against the work with sufficient pressure to cause penetration
C. a cutter being pushed or pulled across the work
D. All of the above.
E. None of the above.

8. List the names of at least three machining operations that have *not* evolved from the lathe.
Any three of the following: electrical discharge machining (EDM), electrochemical machining (ECM), chemical milling, chemical blanking, hydrodynamic machining (HDM), ultrasonic machining, electron beam machining (EBM), laser machining.

9. The control system designed by MIT operated on instructions coded in the _____ system.
binary number

microchip 10. The introduction of the _____ made the use of onboard computers on individual machine tools possible.

D 11. CNC machine tools _____.
A. use computer programs
B. have menu-selectable displays
C. are easier to use than manually controlled machines
D. All of the above.
E. None of the above.

robotic 12. The use of _____ systems for loading and unloading permits some machine tools to operate unattended during the entire machining cycle.

Name ________________________________

13. List at least three of the decisions that a machinist must make before manufacturing a part.
Evaluate individually. Answers may include any three of the following: make a thorough study of the print; determine the machining that must be done; ascertain tolerance requirements; plan the machining sequence; determine how the setup will be made; select the machine tool, cutter(s), and other tools and equipment that will be needed; calculate cutting speeds and feeds; select the proper cutting fluid for the material being machined.

14. The National Institute for Metalworking Skills (NIMS) uses _____ as the basis for industry-recognized certification through performance testing.
skill standards

15. Since machine tools are required to manufacture machine tools, how could there be early machine tools when there were no machine tools to make them? For example, how were the first screw threads cut? List four or five of your own ideas on how the design and development of the first machine tools came about.
Evaluate individually.

Notes

CHAPTER 2

Careers in Machining Technology

Name ______________________ Date __________ Class __________

Learning Objectives

After studying this chapter, you will be able to:

- List the requirements for the various machining technology occupations.
- Explain where to obtain information on occupations in machining technology.
- State what industry expects of an employee.
- Describe what an employee should expect from industry.
- Summarize the information given on a résumé.

Carefully read the chapter, then answer the following questions in the space provided.

B 1. Semiskilled workers are those who perform operations that _____.

A. require at least four years of training
B. do *not* require a high degree of skill or training
C. require an associate's degree
D. All of the above.
E. None of the above.

2. There is little chance for advancement from semiskilled jobs without _____.
additional study, training

3. A common method of becoming a *skilled* worker is through a(n) _____ program.
apprenticeship

four, 4 4. An apprentice will study under a skilled machinist for _____ or more years.

D 5. Many skilled workers entering the field today received their training in _____.
A. community college
B. the armed forces
C. high school
D. All of the above.
E. None of the above.

6. Give a brief description of the work done by a *toolmaker*.
Specializes in producing tools, dies, and fixtures that are necessary for modern mass-production techniques.

7. Give a brief description of the work done by a *diemaker*.
Toolmaker who specializes in making the punches and dies needed to stamp out such parts as auto body panels and electrical components. Also produces dies for making extrusions and die castings.

8. Give a brief description of the work done by a *setup specialist*.
Person who locates and positions tooling and work-holding devices on a machine tool for use by a machine tool operator. This worker may also show the machine tool operator how to do the job, and often checks the accuracy of the machined part.

9. Give a brief description of the work done by the *supervisor* or *manager*.
Usually a skilled machinist who has been promoted to a position of greater responsibility. This person will direct other workers in the shop and is responsible for meeting production deadlines and keeping work quality high. In many shops, the manager may also be responsible for training and other tasks.

10. Give a brief description of the work done by a *parts programmer*.
Inputs data into a computer-controlled (CNC) machine tool for machining a product by determining the sequences, tools, and motions the machine must carry out to complete the machining operation.

D 11. The technician _____.
A. must have a broad educational background
B. operates between the shop and engineering department
C. is a member of a production team
D. All of the above.
E. None of the above.

Name ____

12. List four engineering disciplines involved in the fields of machining, metalworking, and manufacturing.
Industrial, mechanical, tool and manufacturing, and metallurgical engineering.

13. Machining technology is a technical area constantly developing new ideas, materials, processes, and manufacturing techniques. List three sources where up-to-date information can be found on careers in machining technology.
Evaluate individually. Answers may include any three of the following: school career center, technical education instructor, the Internet, state employment services, trade unions, U.S. Department of Labor, community colleges.

14. Employers look for many things in potential employees. List four of them.
Evaluate individually. Answers may include any four of the following: Skills and knowledge, integrity and honesty, comprehension, dependability, teamwork, communication, self-confidence, accountability, initiative, and proper grooming and dress.

E 15. A résumé is a summary of your education and employment background, and should include _____.
   A. your full name and contact information
   B. education and special training
   C. a list of previous employers
   D. a list of references
   E. All of the above.

16. Termination from a job or failure to receive a promotion can be attributed to many factors. List four of them.
Evaluate individually. Answers may include any four of the following: alcohol or illegal drug use on the job, inability or refusal to perform required work, habitual tardiness or missing work repeatedly, inability to work with coworkers, fighting or making threats, inability to work as a team.

Additional Activities

17. Contact several employers of machinists. Secure copies of their job applications, employment opportunities, and job descriptions. Evaluate and summarize the information. Be prepared to discuss the material you have collected with the class. (It might be helpful if several students/trainees worked together gathering the information so there will be no repetition on sources of material. Each can summarize the information they have gathered.)

    Note: If there are no industries employing machinists in your immediate area, contact other employers for job applications and related occupational information.

    Summary:
    Additional activities 17–20 involve instructor's permission and/or group participation. Review activities and assign as applicable or as time permits.

18. As a class project (with instructor's approval), invite a member of the state employment office or the employment officer of a firm employing machinists to speak to your training class. Request that they discuss the following:
    A. The qualities or traits that are sought when hiring a person.
    B. What can discourage them when meeting a potential employee.
    C. The most common causes for a person *not* being hired.
    D. What type of encouragement (paid training, advanced study programs) is offered to employees to advance in the company.

    Summarize the discussion below. Be prepared to discuss the interview in class.
    Evaluate individually.

Name ______________________________

19. *What should an employee expect from an employer*? Since industry expects certain standards from the people it employs, industry should have a responsibility to its employees. What do you think these responsibilities should be?
    Evaluate individually.

20. When meeting a person for the first time at the interview, what would discourage *you* from hiring them?
    Evaluate individually.

21. The following job application has been provided to help you gather the general information required on most job applications. Fill it out as completely as you can, making sure the information is correct.
    The completed job application should be evaluated individually.

APPLICATION FOR EMPLOYMENT

Please print all information. You must fully and accurately complete the application.

PERSONAL INFORMATION

Date

Name
Last First Middle

Present Address
Street City State

Permanent Address
Street City State

Phone No. Alternate No.

If related to anyone in our employ, state name and department Referred by

EMPLOYMENT DESIRED

Position Date you can start Salary desired

Are you employed now? If so may we inquire of your present employer?

Ever applied to this company before? Where When

EDUCATION

	Name and Location of School	Years Compeleted	Subjects Studied	Degree Earned
Grammer School				
High School				
College/University				
Trade, Business or Correspondnece School				

Subject of special study or research work

What foreign languages do you speak fluently? Read fluently? Write fluently?

U.S. Military Service Rank Present membership in National Guard or Reserves

Name ____

Activities other than religious (Exclude organizations the name or character of which indicates the race, creed, color or national origin of its members.)

FORMER EMPLOYERS List below last three employers starting with last one first

Date Month and Year	Name and Address of Employer	Salary	Position	Reason for Leaving
From ______ To				
From ______ To				
From ______ To				

REFERENCES Give below the names of two persons not related to you whom you have known for at least one year

	Name	Address	Job Title	Years Acquainted
1				
2				

PHYSICAL RECORD

Have you any disabilities that might affect your ability to perform this job? ______________________

____

In case of emergency notify ____

Name Address Phone No.

We are an equal opportunity employer. We are dedicated to a policy of non-discrimination in employment on any basis including race, creed, age, sex, religion, national origin, height, weight, marital status, or disability.

I understand that to accept employment, I must be lawfully authorized to work in the United States, and I must present documents to prove my eligibility.

I understand that the company may thoroughly investigate my work and personal history and verify all data given on this application, on related papers, and in interviews. I authorize all individuals, schools, and firms named therein, except my current employer if so noted, to provide any information requested about me, and I release them from all liability for damage in providing this information.

The information on the application and any statement made in conjunction with this application is correct and true to the best of my knowledge. I understand that any false or misleading statement made by me in connection with this application or the failure to disclose any material will be grounds for immediate dismissal.

In consideration of my employment, I agree to conform to the rules and regulations of this company, and my employment and compensation can be terminated, with or without notice, at any time, at the option of either the company or myself. I understand that no manager or representative of the company, other than the president/owner of the company, has any authority to enter into any agreement for employment for any specified period of time, or to make any agreement contrary to the foregoing.

I authorize investigation of all statements contained in this application. I understand that misrepresentation or omission of facts called for is cause for dismissal.

Date Signature ____

Notes

CHAPTER 3

Shop Safety

Name ______________________ Date ____________ Class ____________

Learning Objectives

After studying this chapter, you will be able to:

- Explain why it is important to develop safe work habits.
- Dress in the proper safety equipment and clothing for a machine shop.
- Recognize and correct unsafe work practices.
- Apply safe work practices when employed in a machine shop.
- Select the appropriate fire extinguisher for a particular type of fire.

Carefully study the chapter, then answer the following questions.

____habit____ 1. Safe work habits should become a force of _____.

__D__ 2. Avoid using compressed air to clean chips from machine tools because _____.

A. it does the job too quickly
B. flying chips may cause eye injuries
C. it could create a dangerous mist that can be injurious to your health
D. Both B and C.
E. None of the above.

__C__ 3. Oily rags must be placed in an approved safety container because _____.

A. the soiled rags will be easier to collect for cleaning
B. personal lockers will be cleaner
C. spontaneous combustion may result
D. All of the above.
E. None of the above.

D 4. When handling long sections of metal stock, you should _____.

A. avoid contact with overhead lights

B. be careful not to bend it

C. check to be sure you will not hit someone with it

D. All of the above.

E. None of the above.

5. How should you dress when working on machine tools?

Evaluate individually. Answers may include a snug-fitting shop coat or apron, long hair tied up, and avoid wearing loose clothing or jewelry.

6. When airborne particles are a hazard, have adequate _____ and wear a(n) _____.

ventilation, dust mask

7. When is it important to wear hearing protection?

When working in a noisy area.

8. When is it important to wear disposable plastic gloves?

When handling solvents, cutting fluids, and oils.

9. When is it important to wear a dust mask?

When machining produces airborne dust particles.

A 10. Approved eye protection should be worn _____.

A. at all times when in the shop

B. only when working with machine tools

C. only when using hand tools

D. None of the above.

D 11. Awareness barriers _____.

A. remind the operator of an area that is dangerous

B. stop the machine when a light beam or electronic beam is broken by someone entering the danger area

C. can be red or yellow lines painted on the floor

D. All of the above.

E. None of the above.

C 12. *Never* attempt to remove chips or cuttings _____.

A. with your hands

B. while the machine is operating

C. Both A and B.

D. None of the above.

Name ________________________________

13. When removing chips or cuttings from a machine, you should use a(n) _____ to clear the machine and _____ to remove long stringy cuttings.
brush, pliers

14. List two times when it is important that you *do not* operate machine tools.
Any two of the following: when your senses are affected by medication, all guards and safety features are not in place, you have not been instructed in the safe operation of the machine.

D 15. Before operating a machine tool, you should _____.
  A. make sure all safety guards are in place
  B. think before acting
  C. remove rings and other jewelry
  D. All of the above.
  E. None of the above.

Match the class of fire listed on the right with the types of combustible materials listed on the left.

C 16. Electrical equipment fires require nonconducting extinguishing agents that will smother the flames.

B 17. Flammable liquid and grease fires require the blanketing or smothering effects of dry chemicals or carbon dioxide.

A 18. Those involving ordinary combustible materials (paper, wood, textiles). They require the cooling and quenching effect of water, or solutions containing a large percentage of water.

D 19. Extinguishers containing specially prepared heat-absorbing dry powder are used on flammable metals, such as magnesium and lithium.

A. Class A fires
B. Class B fires
C. Class C fires
D. Class D fires

20. What plans have you made in the event of a fire in the training area?
Evaluate individually. Answers may include knowing where the fire exits are, knowing alternate escape routes, and knowing the location of fire extinguishers.

21. In the space provided below, list four or five unsafe work habits you have observed in the shop area. Be prepared to discuss them in class. You may be able to prevent a serious injury. Do *not* include individual's names.
Evaluate individually.

22. In the space provided below, list situations or conditions that you feel are unsafe or not suitably protected to prevent possible injury. Be prepared to discuss how they can be corrected.
    Evaluate individually.

23. In the space provided below, list four or five safe work habits you have observed in the shop area. Be prepared to discuss them in class. Do *not* include individual's names.
    Evaluate individually.

CHAPTER 4
Understanding Drawings

Name ______________________ Date __________ Class __________

Learning Objectives

After studying this chapter, you will be able to:

- Read drawings that are dimensioned in fractional inches, decimal inches, and in metric units.
- Explain the information found on a typical drawing.
- Describe how detail, subassembly, and assembly drawings differ.
- Indicate why drawings are numbered.
- Explain the basics of geometric dimensioning and tolerancing.

Carefully study the chapter, then answer the following questions.

___D___ 1. Drawings are used by industry to _____.

A. show what an object looks like in multiview
B. show the machinist what to make
C. show the machinist which standards to follow
D. All of the above.
E. None of the above.

___fractional___ 2. Drawings using _____ dimensioning usually show objects that do not require a high degree of precision in their manufacture.

3. _____ is a system that employs English and metric dimensions on the same drawing.
___Dual dimensioning___

4. List at least four types of information (besides dimensions), that is included on a drawing.
Evaluate individually. Answers may include any four of the following: material(s) to be used; surface finish required; tolerances; quantity of units per assembly; scale of drawing; next assembly or subassembly; revisions; the name of the object.

5. Define *tolerance*.
Tolerances are allowances, either undersize or oversize, permitted when machining or making an object.

6. A machinist can compare some surface finishes with required specifications by employing a surface _____ standard.
roughness comparison

profilometer 7. If the surface is critical, the surface finish is measured electronically with a device called a _____.

D 8. Dimensions should be scaled off of a drawing _____.
   A. when there isn't enough room on the print
   B. when using computer aided design (CAD) software
   C. when doing revisions
   D. Dimensions should *never* be scaled off a drawing.

B 9. A subassembly drawing differs from an assembly drawing by showing _____ of complete object or project.
   A. several small portions
   B. only a small portion
   C. only the front and back portions
   D. None of the above.

D 10. Standard size drawings have been developed to _____.
   A. simplify the storage of drawings
   B. make prints easier to handle
   C. make prints easier to stock
   D. All of the above.
   E. None of the above.

Name ____

11. Prepare a *Parts List and Bill of Materials* for the object shown in the figure below.
Evaluate individually.

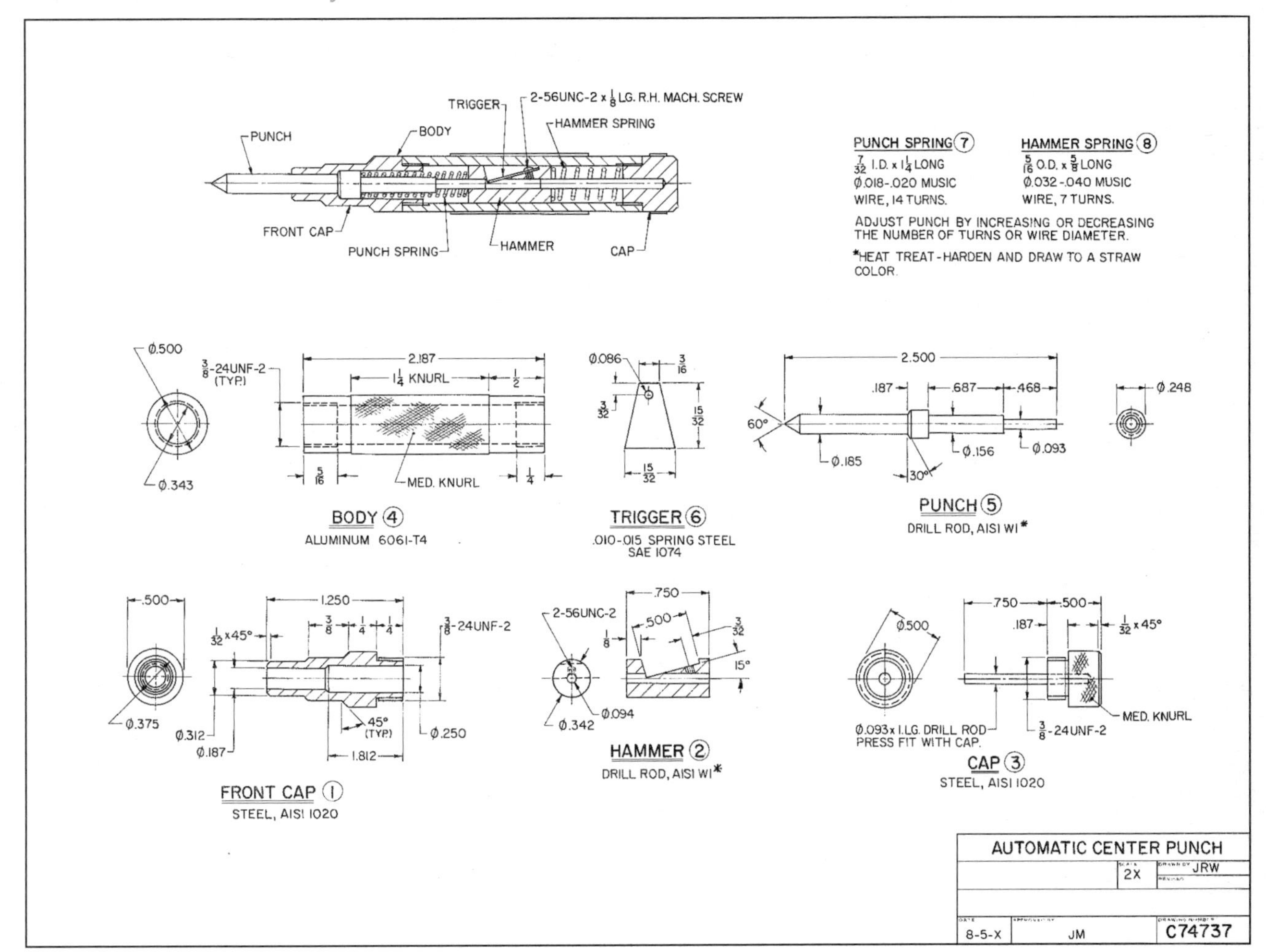

12. Why was geometric dimensioning and tolerancing developed?
When the amount of variation (tolerances) in form (shape and size) and position (location) needs to be more strictly defined, it provides the precision needed to allow for the most economical manufacture of parts.

13. What is the geometric dimensioning and tolerancing system?
Geometric dimensioning and tolerancing is a system that provides additional precision compared to conventional dimensioning. It ensures that parts can be easily interchanged.

14. Why are geometric characteristic symbols employed?
They are employed to provide clarity and precision in communicating design specifications.

___D___ 15. *Geometric characteristic symbols* indicate tolerances of _____. They indicate the total amount a specific dimension is permitted to vary from design specifications.

A. form and profile
B. runout
C. location and orientation
D. All of the above.
E. None of the above.

___B___ 16. A(n) _____ is a numerical value denoting the exact size, profile, orientation, or location of a feature.

A. reference dimension
B. basic dimension
C. LMC
D. None of the above.

___A___ 17. A(n) _____ is used for information only. It is not used for production or inspection purposes.

A. reference dimension
B. basic dimension
C. LMC
D. None of the above.

18. _____ condition would indicate minimum hole diameter or maximum shaft size.
Maximum material

19. _____ condition would indicate maximum hole diameter or minimum shaft diameter.
Least material

20. What is *actual size*?
The measured size of a part after it is manufactured.

21. A _____ contains the geometric symbol, allowable tolerance, and the datum reference letter(s). It is connected to an extension line of the feature.
feature control frame

Name ________________________________

Refer to the drawing below to answer questions 22 and 23.

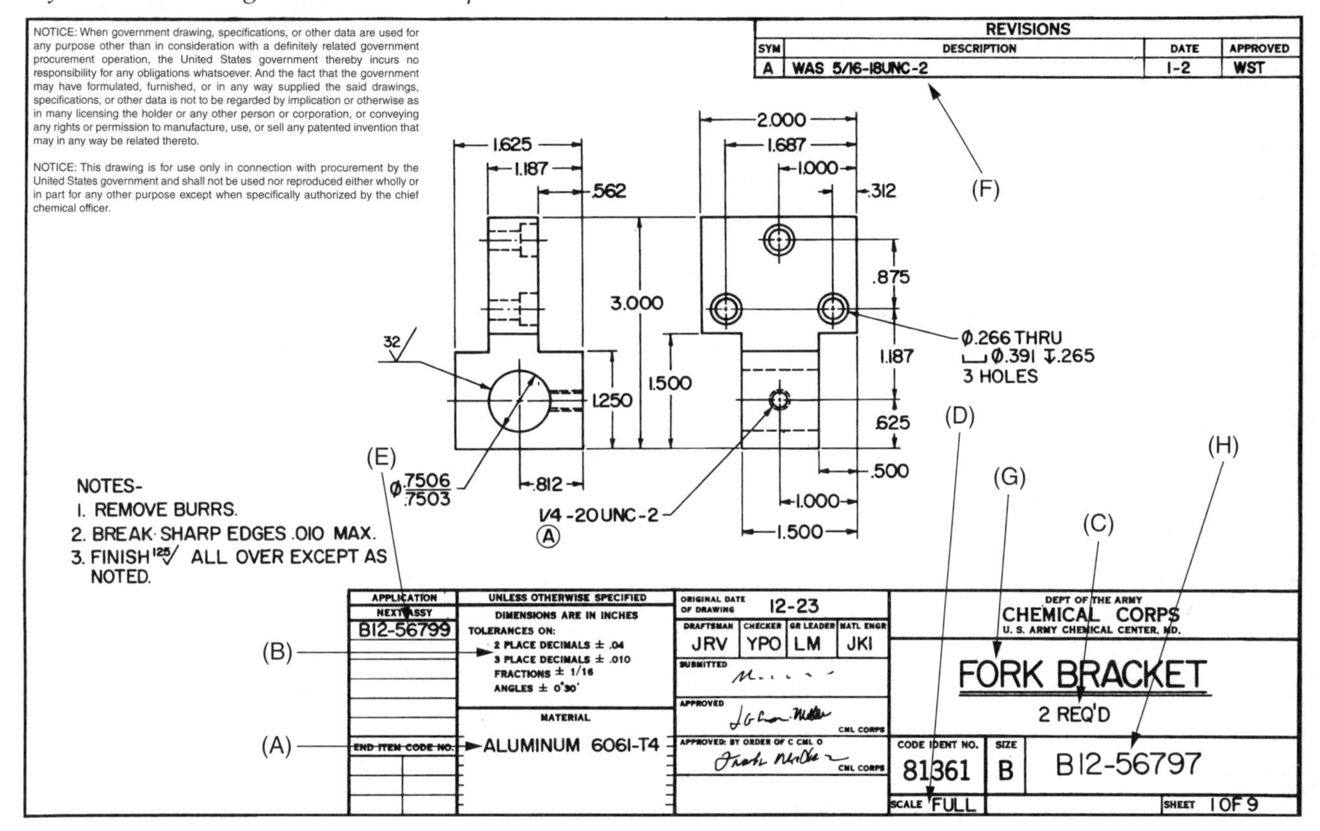

22. Identify the parts indicated.

A. Material to be used

B. Tolerances

C. Quantity

D. Scale

E. Next assembly

F. Revisions

G. Name of object

H. Drawing number

23. Answer the following with information from the same drawing used for question 22.

A. Height of part: 3.000″

B. Width of part: 2.000″

C. Thickness of part: 1.625″

D. Diameter of large hole: 0.7503″

E. Tolerance on large hole: +0.0003″

F. Diameter of small holes: 0.266″

G. Counterbore depth on small holes: 0.265″

H. Diameter of counterbore: 0.391″

I. Special NOTES: Remove burrs. Break sharp edges .010″ Max. Finish 125 all over except as noted.

Carefully study the drawing shown below. Use information from the drawing to answer questions 24–29.

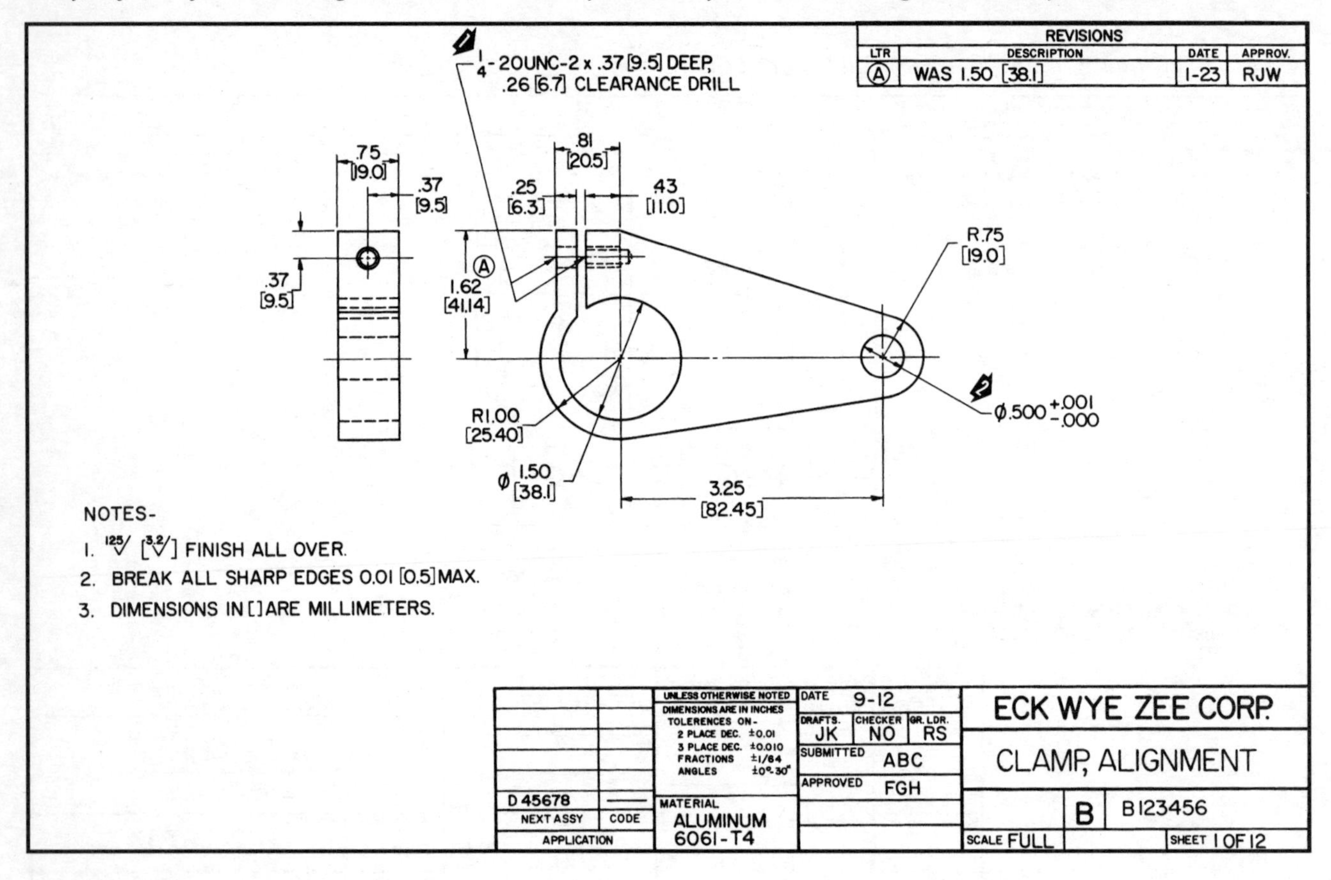

24. Identify the information indicated.
    A. Part name: Clamp, Alignment
    B. Drawing scale: Full size
    C. Drawing number: B123456
    D. Next assembly: D45678
    E. Type of dimensioning used: Dual dimensioning
    F. Material specified: Aluminum 6061-T4
    G. Number of drawings making up assembly: 12
    H. What revisions were made: Distance from centerline to flat on top of part was changed from 1.50″ (38.0 mm) to 1.62″ (41.14 mm).

25. Minimum size material needed to make the part. Do *not* add to dimensions for machining.
    L. 4.25″ (107.85 mm)
    W. 2.62″ (66.54 mm)
    T. 0.75″ (19.0 mm)

26. Special NOTES:
    1. 125 [3.2] all over.
    2. Break all sharp edges 0.01 [0.5] MAX.
    3. Dimensions in [ ] are millimeters.

Name ______________________________

27.

A. Diameter of large hole: 1.50″ (38.1 mm)

B. Diameter of small hole: 0.500″

28. No metric dimension is given for machining the small hole. Why not?
No standard metric tool available this size.

29.

A. Distance between hole centerlines: 3.25″ (82.45 mm)

B. Thread size: 1/4-20UNC-2

C. Why isn't a metric thread size given? There is no metric thread this size.

D. Thread depth: 0.37″ (9.5 mm)

E. Thread clearance drill size: 0.26″ (6.7 mm)

F. Width of slot: 0.13″ (3.5 mm)

G. Thickness of part: 0.75″ (19.0 mm)

H. Tolerances allowed on small hole: +0.001″

I. Radius of part at small end: 0.75″ (19.0 mm)

J. Radius of part at large end: 1.00″ (25.4 mm)

K. How far is flat on top of part located above horizontal centerline? 1.62″ (41.14 mm)

Notes

CHAPTER

5

Measurement

Name ______________________ Date ____________ Class ____________

Learning Objectives

After studying this chapter, you will be able to:

- Measure to 1/64″ (0.5 mm) with a steel rule.
- Measure to 0.0001″ (0.002 mm) using a Vernier micrometer caliper.
- Measure to 0.001″ (0.02 mm) using Vernier measuring tools.
- Measure angles to 0°5′ using a universal Vernier bevel.
- Identify and use various types of gages found in a machine shop.
- Use a dial indicator.
- Use the various helper measuring tools found in a machine shop

Carefully study the chapter, then answer the following questions.

1. Make readings from the ruler shown below.

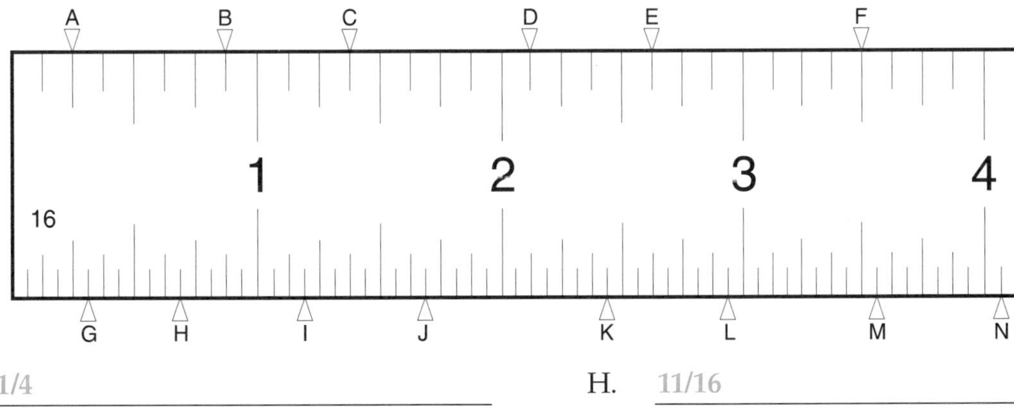

A. 1/4

B. 7/8

C. 1 3/8

D. 2 1/8

E. 2 5/8

F. 3 1/2

G. 5/16

H. 11/16

I. 1 3/16

J. 1 11/16

K. 2 7/16

L. 2 15/16

M. 3 9/16

N. 4 1/16

2. Make readings from the ruler shown below.

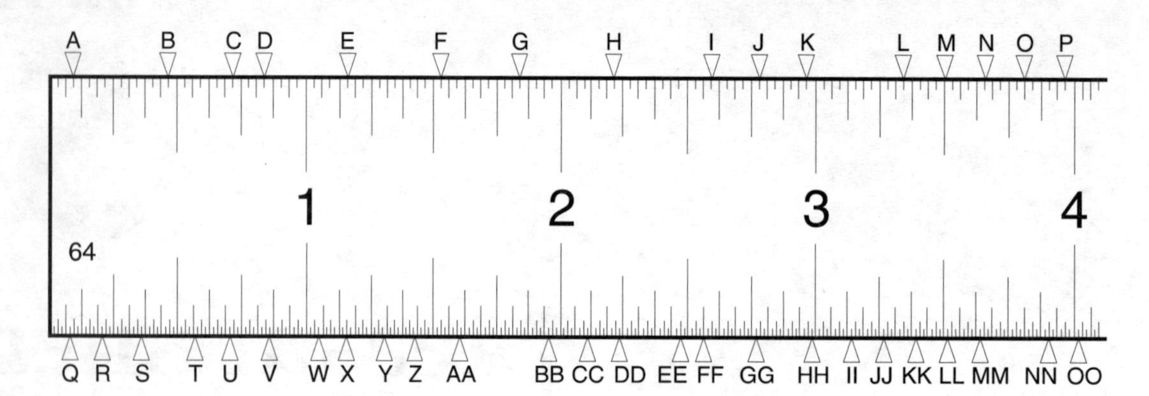

A. 3/32

B. 15/32

C. 23/32

D. 27/32

E. 1 5/32

F. 1 17/32

G. 1 27/32

H. 2 7/32

I. 2 19/32

J. 2 25/32

K. 2 31/32

L. 3 11/32

M. 3 1/5

N. 3 21/32

O. 3 13/16

P. 3 31/32

Q. 5/64

R. 13/64

S. 23/64

T. 37/64

U. 45/64

V. 55/64

W. 1 3/64

X. 1 5/32

Y. 1 19/64

Z. 1 27/64

AA. 1 39/64

BB. 1 61/64

CC. 2 7/64

DD. 2 15/64

EE. 2 31/64

FF. 2 39/64

GG. 2 49/64

HH. 2 63/64

II. 3 9/64

JJ. 3 17/64

KK. 3 25/64

LL. 3 33/64

MM. 3 41/64

NN. 3 29/32

OO. 4 1/64

Name ____

3. Make the readings from the metric ruler shown below.

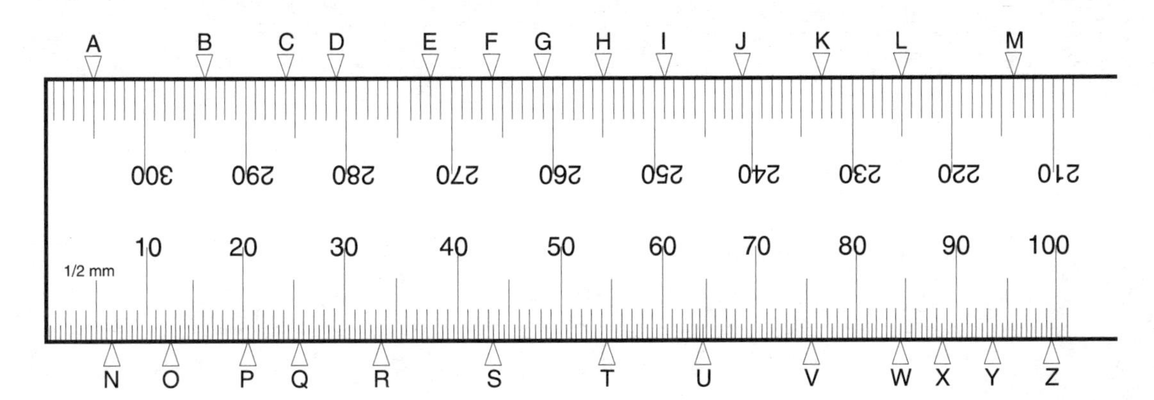

A. 305.0

B. 294.0

C. 286.0

D. 281.0

E. 272.0

F. 266.0

G. 261.0

H. 255.0

I. 249.0

J. 241.0

K. 233.0

L. 225.0

M. 214.0

N. 6.5

O. 12.5

P. 20.5

Q. 25.5

R. 33.5

S. 33.5

T. 43.5

U. 64.5

V. 75.5

W. 84.5

X. 88.5

Y. 93.5

Z. 99.5

4. Make readings from the micrometers shown below.

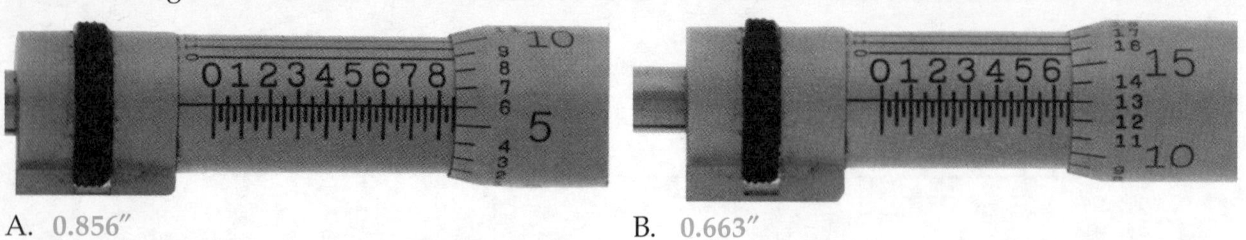

A. 0.856″ B. 0.663″

5. Make readings from the micrometers shown below.

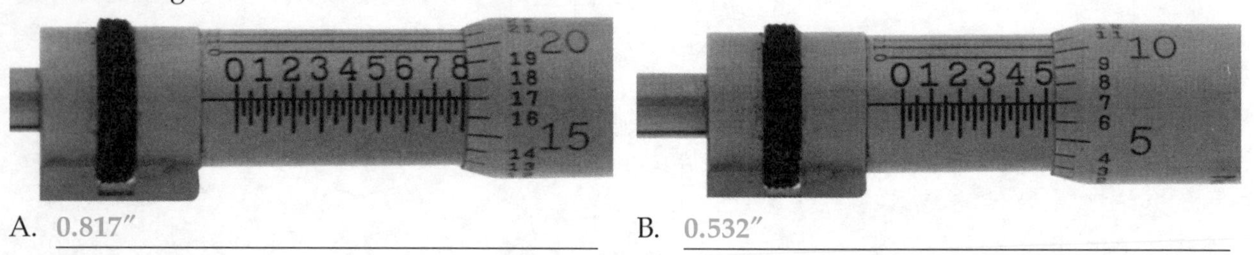

A. 0.817″ B. 0.532″

6. Make readings from the micrometers shown below.

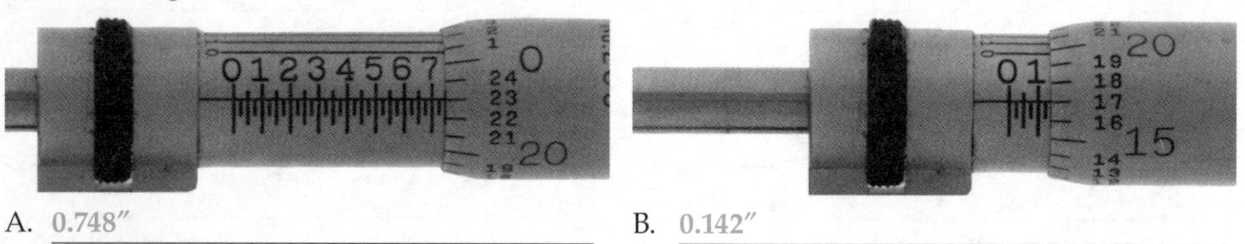

A. 0.748″ B. 0.142″

7. Make readings from the micrometers shown below.

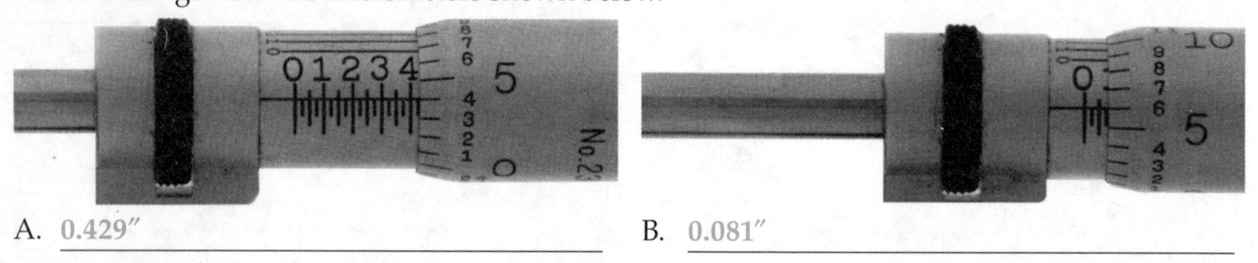

A. 0.429″ B. 0.081″

8. Make readings from the micrometers shown below.

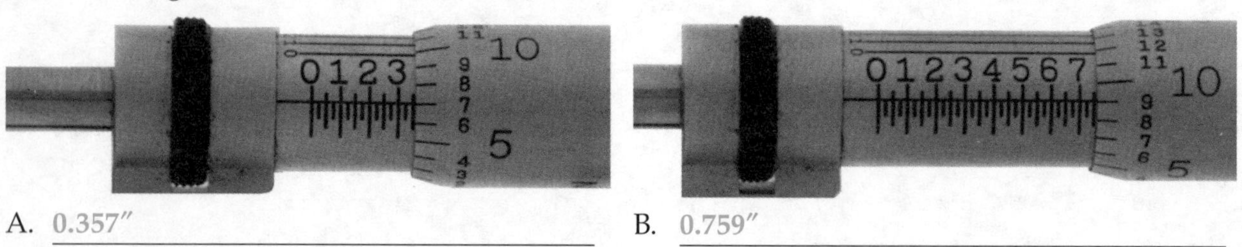

A. 0.357″ B. 0.759″

9. Make readings from the micrometers shown below.

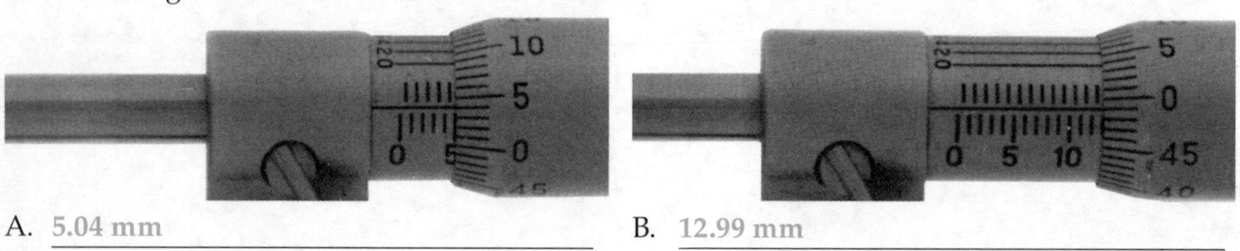

A. 5.04 mm B. 12.99 mm

Name ______________________________

10. Make readings from the micrometers shown below.

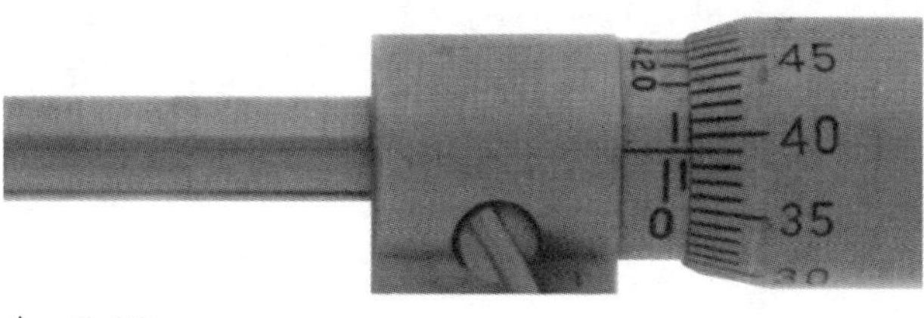

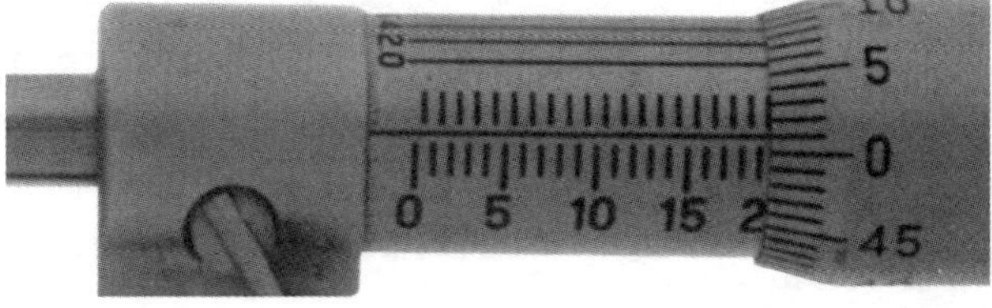

A. 1.39 mm ______________________________ B. 19.51 mm ______________________________

11. Make readings from the micrometers shown below.

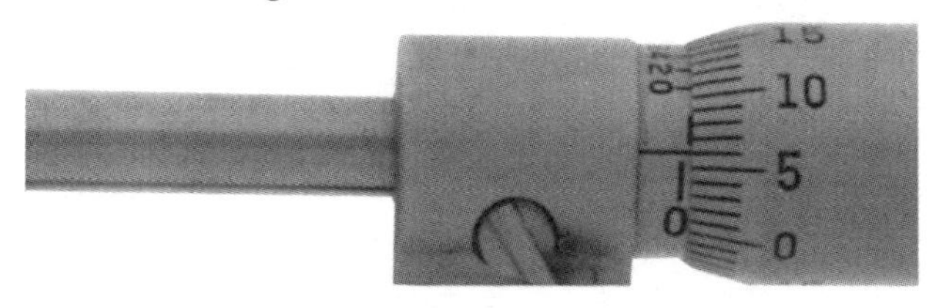

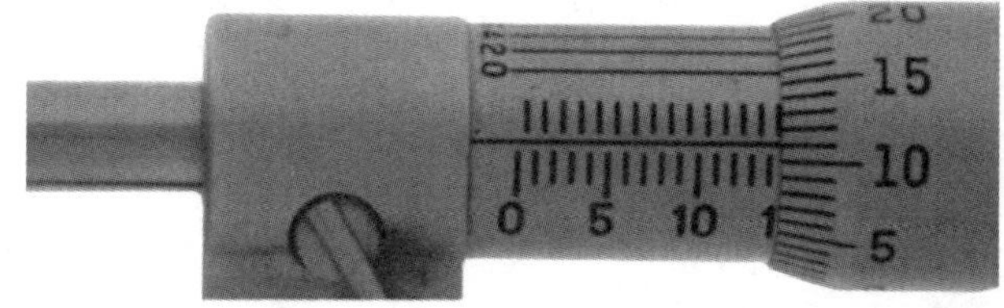

A. 0.56 mm ______________________________ B. 14.61 mm ______________________________

12. Make readings from the micrometers shown below.

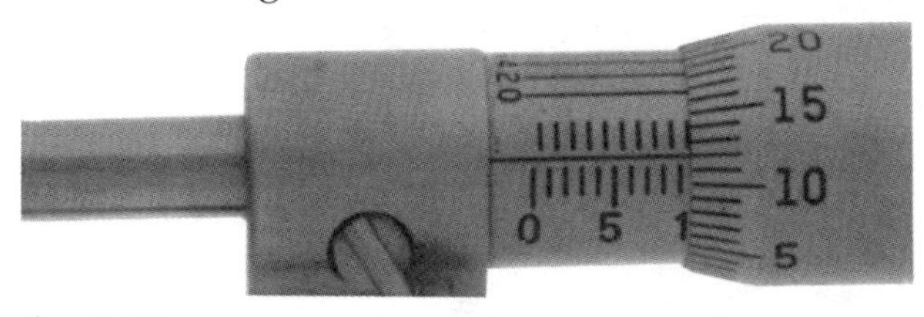

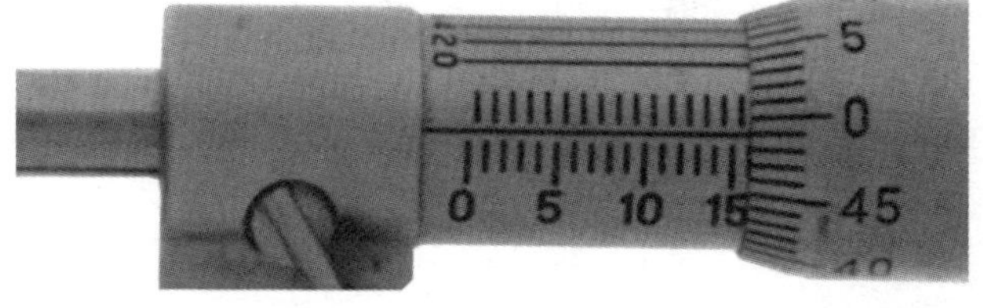

A. 9.62 mm ______________________________ B. 15.99 mm ______________________________

13. Make readings from the micrometers shown below.

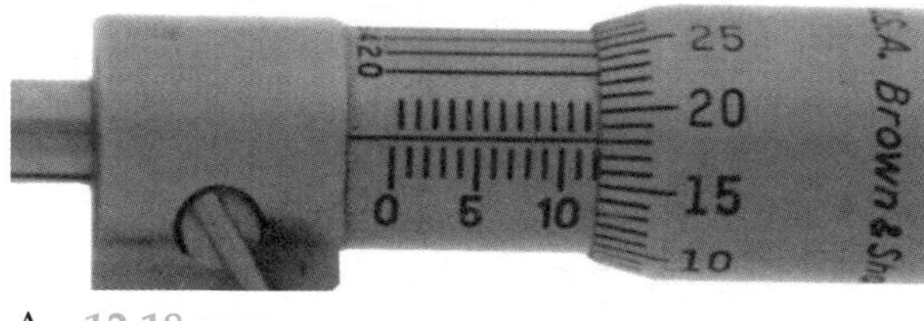

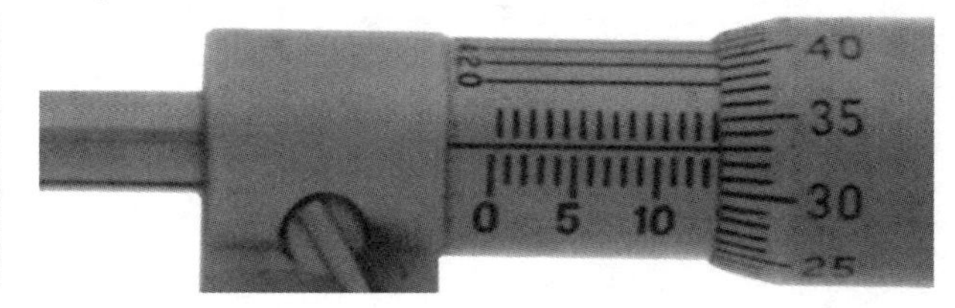

A. 12.18 mm ______________________________ B. 13.83 mm ______________________________

14. Make readings of the Vernier calipers shown below.

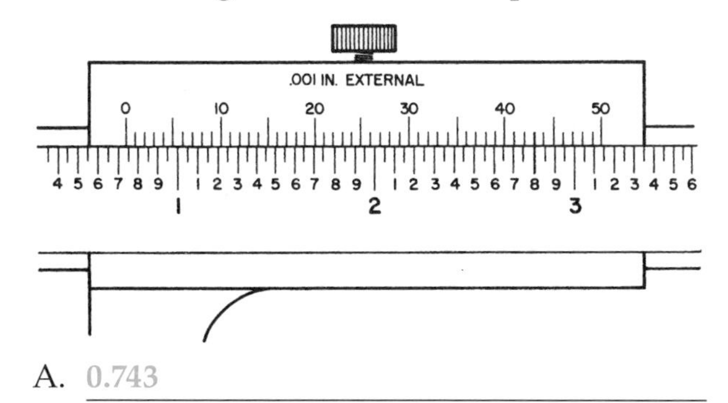

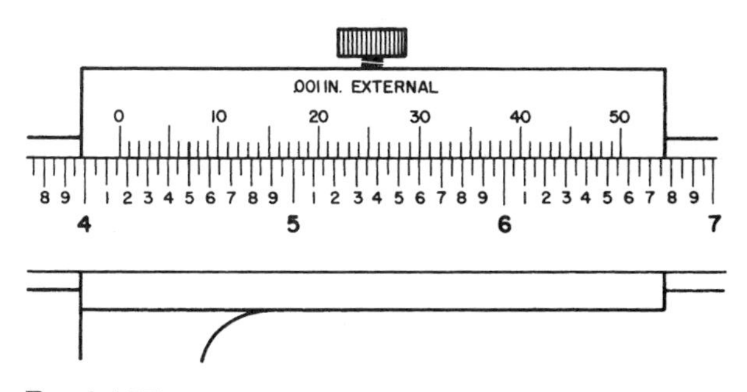

A. 0.743 ______________________________ B. 4.157 ______________________________

15. Make readings of the Vernier calipers shown below.

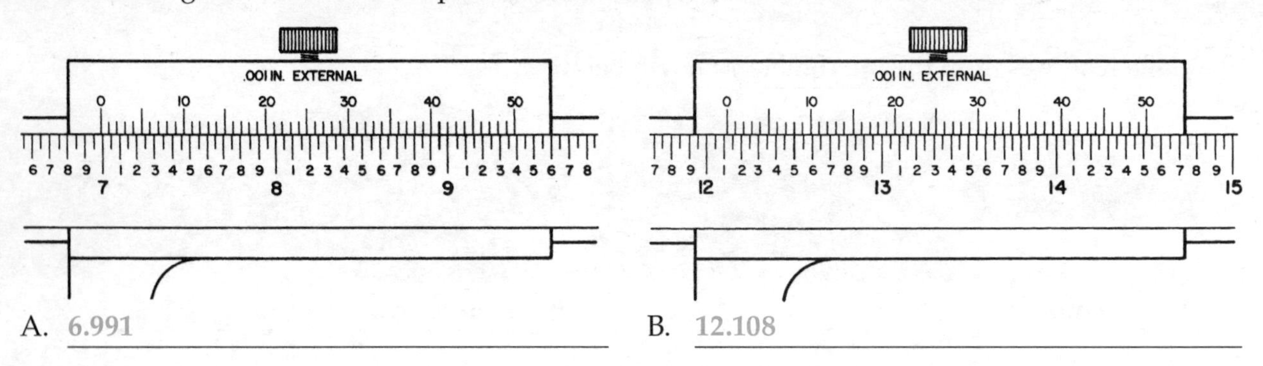

A. 6.991 B. 12.108

16. Make readings of the Vernier calipers shown below.

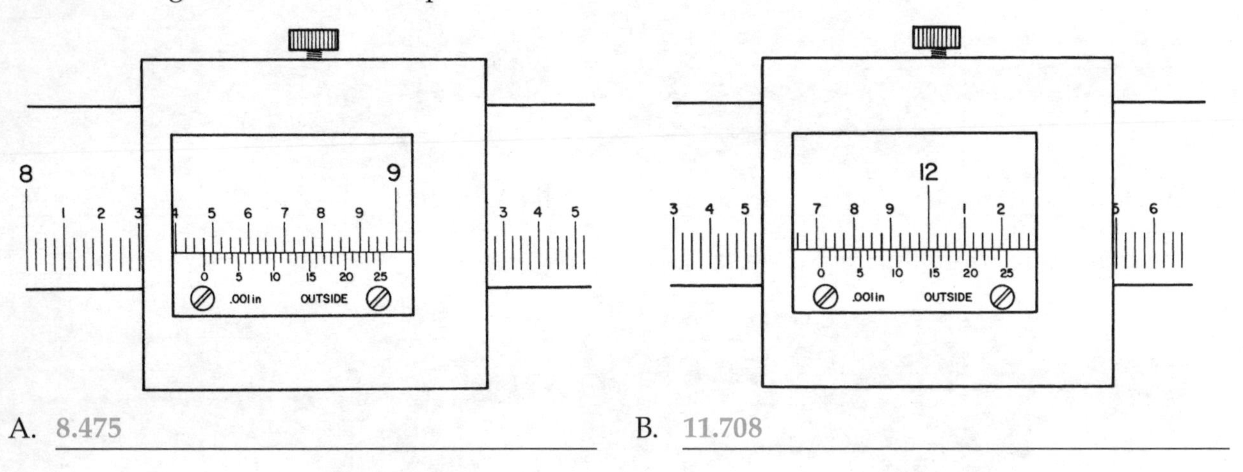

A. 8.475 B. 11.708

17. Make readings of the Vernier calipers shown below.

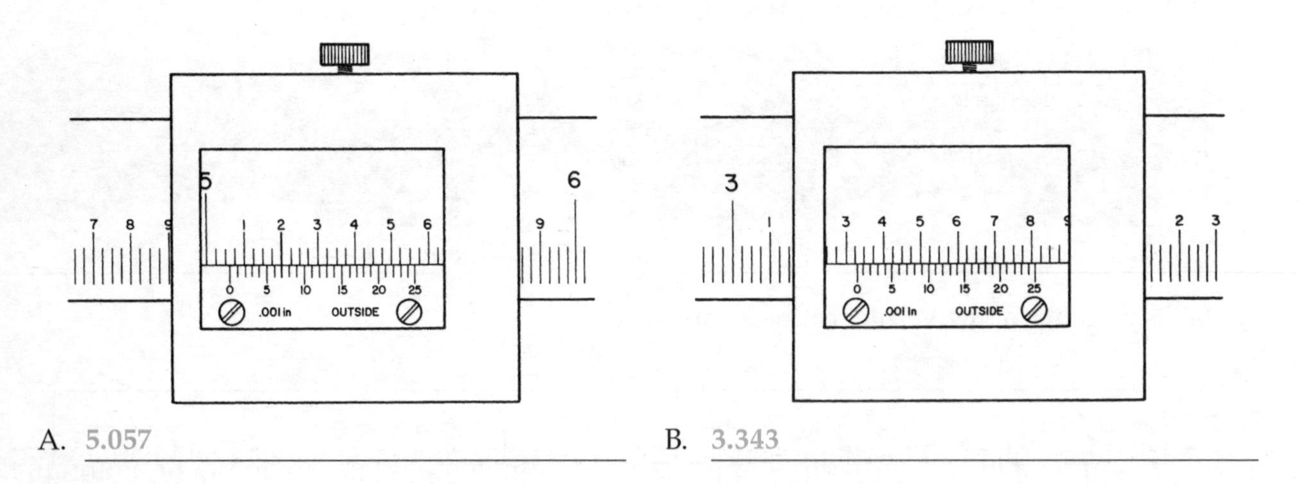

A. 5.057 B. 3.343

Name ______________________________

18. Make readings of the Vernier calipers shown below.

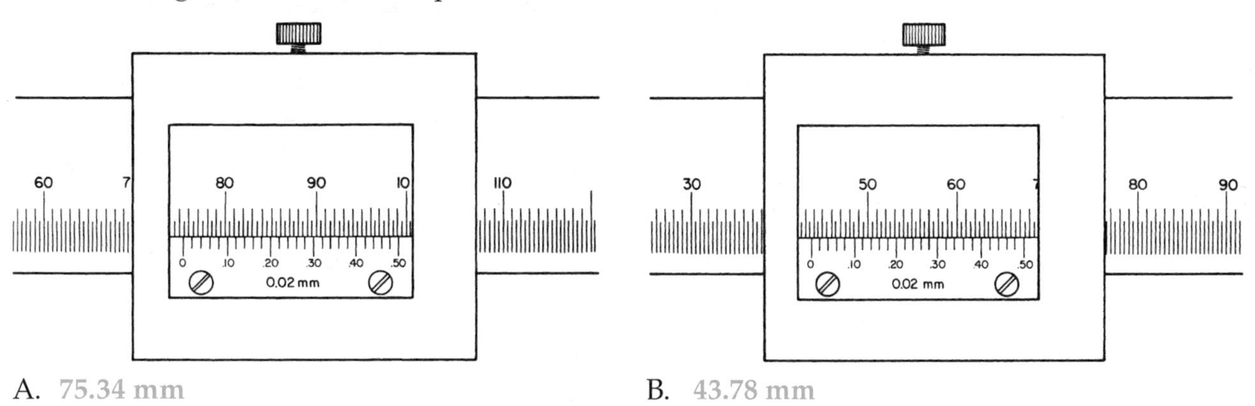

A. 75.34 mm

B. 43.78 mm

19. Make readings of the Vernier calipers shown below.

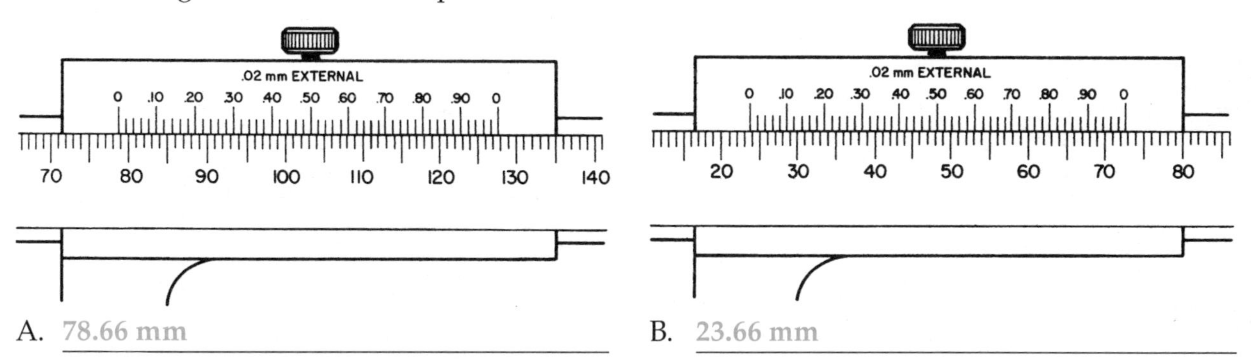

A. 78.66 mm

B. 23.66 mm

20. Make readings of the Vernier calipers shown below.

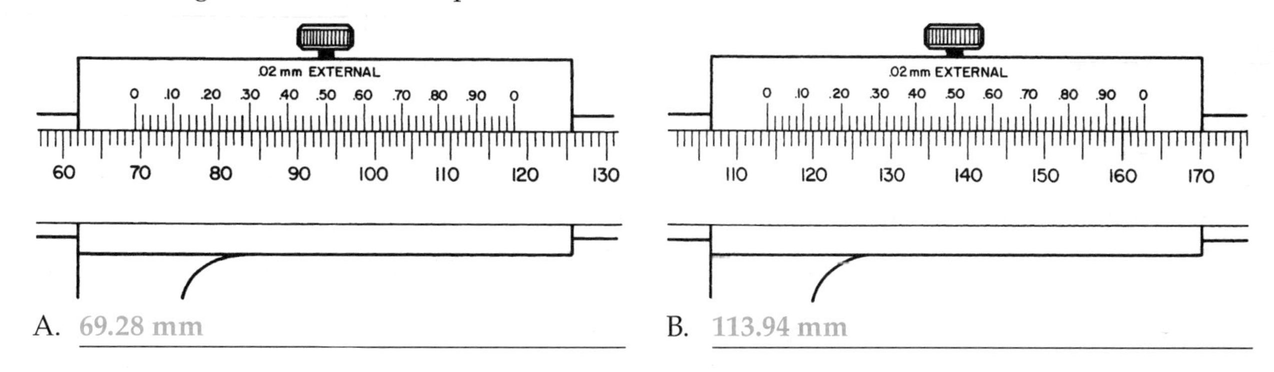

A. 69.28 mm

B. 113.94 mm

21. The micrometer caliper, known as a mike, is a precision measuring tool capable of reading to _____ of an inch, and when fitted with a Vernier scale to _____ of an inch.
0.001, 0.0001

22. A metric micrometer caliper is capable of reading to the _____ part of a millimeter, and to _____ when fitted with a Vernier scale.
0.01 mm, 0.002 mm

D 23. A Vernier caliper has which of the following advantages over a micrometer?

A. Easier to use.

B. It can be used to make both inside and outside measurements over a range of sizes.

C. Permits a range of measurements that would require several sizes of micrometers.

D. Both B and C.

E. None of the above.

24. How does a dial caliper differ from a Vernier caliper?

Evaluate individually. Answers may include that the dial caliper is used for direct reading and that it can be locked for repetitive measurements.

25. The universal Vernier protractor can measure angles accurately to _____ of a degree or _____ minutes.

1/12, 5

26. The micrometer caliper and Vernier caliper are precision measuring tools that require care in use if they are to retain their accuracy. List five precautions that must be observed when using them.

Evaluate individually. Answers may include any four of the following: Place the micrometer on the work carefully so the faces of the anvil and spindle will not be damaged. Clean the anvil and spindle faces before use. Avoid "springing" a micrometer by applying too much pressure when you are making a measurement. Keep the micrometer clean. Avoid placing a micrometer where it may fall on the floor or have other tools placed on it. Never attempt to make a micrometer reading until a machine has come to a complete stop. Clean and oil the tool if it is to be stored for some time. If possible, place the micrometer in a small box for protection.

27. Plug gages are used to check whether hole diameters are within specified _____.

tolerances

28. Where are the plugs located on a progressive plug gage?

The plugs are located on the same side of the gage, right next to each other.

D 29. Dial indicators would be classified as _____.

A. helper type measuring tools

B. direct measuring tools

C. a combination of both types

D. All of the above.

E. None of the above.

Name ____

___D___ 30. Dial indicators can be used for _____.

A. centering and aligning work on machine tools
B. checking for eccentricity
C. visual inspection
D. All of the above.
E. None of the above.

31. How does an air gage measure the diameter of a bore?
The air gage measures the air leakage between the plug and the hole wall.

32. What is an *optical comparator*?
A measuring device that makes use of an enlarged image of the part to be inspected. The image is projected on a screen where it is superimposed on an enlarged accurate drawing of the part.

___light waves___ 33. Optical flats employ _____ as a standard to make measurements.

34. The screw pitch gages illustrated below can be used to determine:
the number of threads per inch or mm on a threaded section.

35. Of what use is the fillet and radius gage?
The thin steel blades of a fillet and radius gage are used to check concave and convex radii on corners or against shoulders. The gage is used for layout work and inspection, and as a template when grinding form cutting tools.

___D___ 36. Drill rods are steel rods used to inspect hole _____.

A. diameter
B. location
C. alignment
D. All of the above.
E. None of the above.

37. What are helper measuring tools?
Measuring devices that are not direct reading but require the aid of a rule, micrometer, or Vernier caliper to determine size of measurement taken.

38. List four helper type measuring tools.
Inside caliper, telescoping gage, outside caliper, and small hole gage.

39. The _____ illustrated below is a tool that can be employed to make accurate internal measurements, but it must be used with a micrometer or Vernier caliper.
telescoping gage

C 40. A small hole gage is used to measure _____.
A. concave and convex radii on corners
B. openings that are too large for a telescoping gage
C. openings that are too small for a telescoping gage
D. All of the above.
E. None of the above.

CHAPTER
6
Layout Work

Name ______________________ Date __________ Class __________

Learning Objectives

After studying this chapter, you will be able to:

- Explain why layouts are needed.
- Identify common layout tools.
- Use layout tools safely.
- Make basic layouts.
- List safety rules for layout work.

Carefully study the chapter, then answer the following questions.

___D___ 1. Layout lines are used to _____.

A. provide the machinist with guidelines for machining
B. show the machinist where to machine
C. eliminate or reduce the possibility of machining incorrectly
D. All of the above.
E. None of the above.

___D___ 2. A good layout job is determined by its _____.

A. neatness
B. accuracy
C. legibility
D. All of the above.
E. None of the above.

___layout dye___ 3. Layout lines are easier to see if a coating of _____ is applied to the metal.

A 4. A(n) _____ is used to lay out circles and arcs.
- A. divider
- B. scriber
- C. ruler
- D. None of the above.
- E. All of the above.

5. Identify the layout tool illustrated below.
Trammel

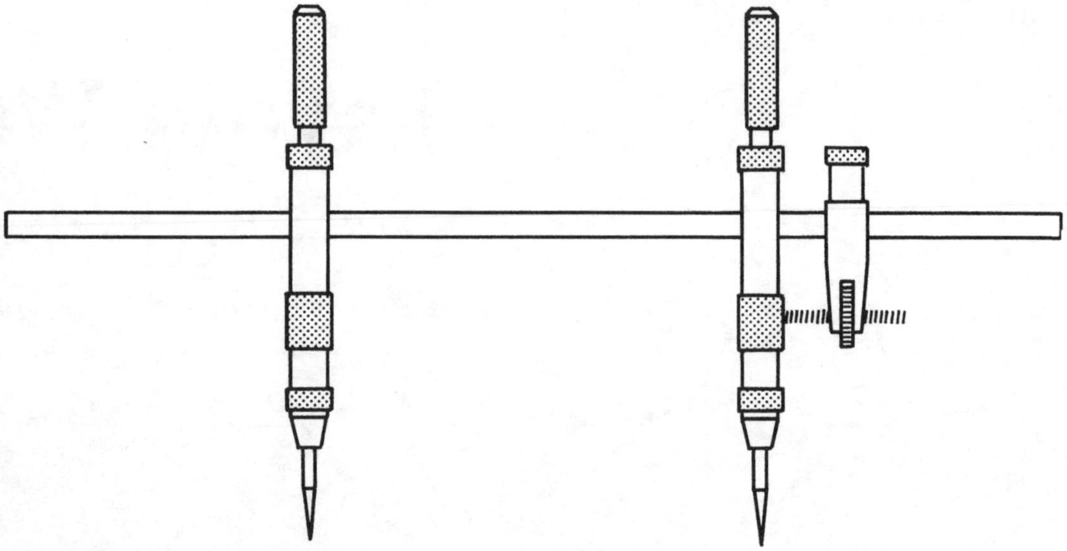

B 6. The hermaphrodite caliper is a layout tool that _____.
- A. can be used to draw lines perpendicular to the edge of the material
- B. can be used to locate the center of irregularly shaped stock
- C. has two legs shaped like a caliper and one shaped like a divider
- D. All of the above.
- E. None of the above.

E 7. A surface gage _____.
- A. consists of a base, spindle, and trammel
- B. is used to scribe lines perpendicular to the surface
- C. cannot be used on a curved surface
- D. All of the above.
- E. None of the above.

B 8. A surface gage can be used to check whether a part is _____ to a given surface.
- A. perpendicular
- B. parallel
- C. equal in width
- D. All of the above.
- E. None of the above.

flatness 9. Surface plate grades are given in degrees of _____.

Name ______________________________

A ______ 10. A(n) _____ is practical for many jobs because the sliding blade is adjustable and interchangeable with other blades.

A. double square
B. protractor depth gage
C. plain protractor
D. All of the above.
E. None of the above.

B ______ 11. A(n) _____ consists of a hardened blade, square head, center head, and bevel protractor.

A. double square
B. combination set
C. hardened steel square
D. All of the above.
E. None of the above.

C ______ 12. A(n) _____ can be used when angles do not need to be laid out or checked to extreme accuracy.

A. universal bevel
B. protractor depth gage
C. plain protractor
D. All of the above.
E. None of the above.

13. Identify the layout tools illustrated below.
Protractor depth gage

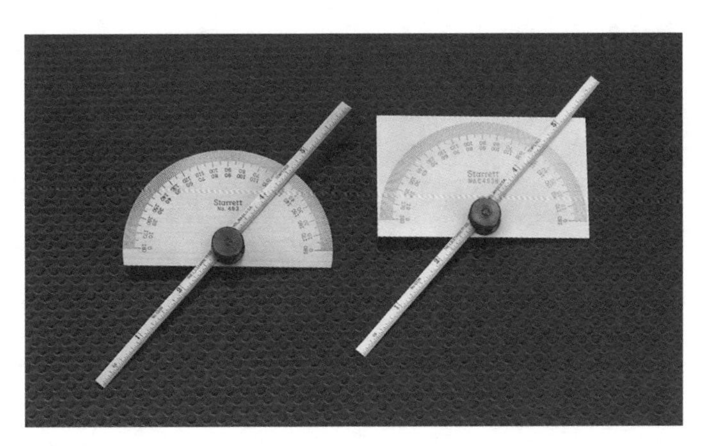

A ______ 14. A(n) _____ is useful for checking, laying out, and transferring angles.

A. universal bevel
B. protractor depth gage
C. plain protractor
D. All of the above.
E. None of the above.

15. Identify the layout tool illustrated below.
 Universal bevel

vernier 16. The ______ protractor must be employed when very accurate angular lines must be laid out.

17. Carefully study the FLANGE drawing shown below. Secure a section of metal that is the proper size and make a layout. Allowable tolerance ± 1/64″.
 Evaluate individually.

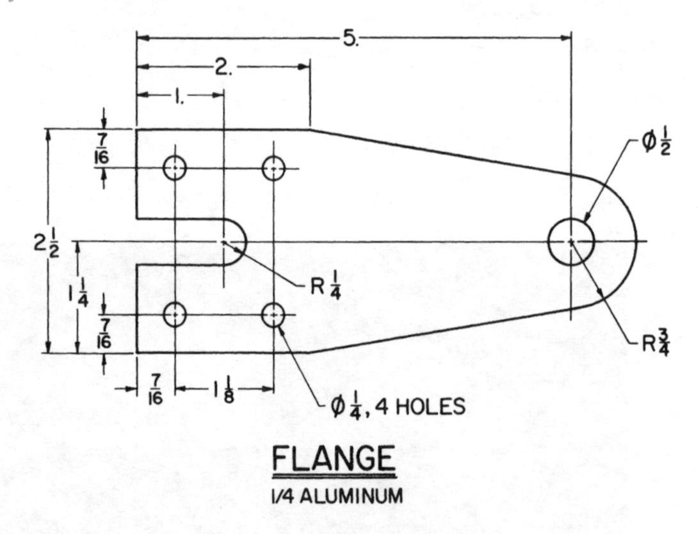

Name ______________________________

18. List the tools needed to lay out the flange in question #18.
    Rule, scribe, square, divider, prick punch, center punch, hammer.

19. In the order they would be done, describe the steps necessary to lay out the flange shown in question #18.
    Evaluate individually. Answers may include: Carefully study the drawing; cut stock to size and remove all burrs and sharp edges; clean all dirt, grease, and oil from the work surface and apply layout dye; locate and scribe a reference line; locate all circle and arc centerlines; punch all points where centerlines intersect; scribe in all circles and arcs; locate and scribe angular lines; connect remaining points.

20. List five safety rules you should follow when doing layout work.
    Any order: never carry an open scriber, divider, trammel, or hermaphrodite caliper in your pocket; always cover sharp points with a cork when the tool is not being used; wear goggles when grinding scriber points; get help when you must move heavy items, such as angle plates or V-blocks; remove all burrs and sharp edges from stock before starting layout work.

Notes

CHAPTER

7

Hand Tools

Name ________________ Date ________ Class ________

Learning Objectives

After studying this chapter, you will be able to:

- Identify the most commonly used machine shop hand tools.
- Select the proper hand tool for the job.
- Maintain hand tools properly.
- Explain how to use hand tools safely.

Carefully read the chapter, then answer the following questions in the space provided.

___D___ 1. A toolmaker's vise _____.
   A. can be rotated to any desired position
   B. can be tilted to any desired position
   C. can hold small precision parts
   D. All of the above.
   E. None of the above.

___A___ 2. When placing work in a vise, clamping action can be obtained by _____.
   A. using the handle to turn the heavy screw
   B. hammering the handle with a soft-face hammer
   C. increasing leverage with an additional length of pipe added to the handle
   D. All of the above.
   E. None of the above.

___paper___ 3. Placing strips of _____ the width of the clamp jaw between the work and the jaws will improve clamping action.

B 4. The plier size indicates the _____.
A. length of the handle
B. overall length of the tool
C. length of the jaws
D. All of the above.
E. None of the above.

E 5. Torque wrenches are used when _____.
A. they fit the bolt or nut better than another wrench
B. a threaded fastener must be tightened to specified limits
C. a threaded fastener must provide maximum holding power without danger of the fastener failing
D. All of the above.
E. Both B and C.

6. When should the handle of a torque wrench be lengthened for additional leverage?
Under no condition should the handle be lengthened for additional leverage.

7. Should an adjustable wrench be pushed or pulled? Why?
Pulled. Pushing any wrench is considered dangerous.

A 8. When using an adjustable wrench, _____.
A. the movable jaw should face the direction the fastener is to be rotated
B. the movable jaw should face the opposite direction that the fastener is to be rotated
C. a hammer can be used to help loosen a stubborn fastener
D. All of the above.
E. None of the above.

C 9. The body or jaw of the _____ completely surrounds the bolt head or nut.
A. spanner wrench
B. pipe wrench
C. box wrench
D. All of the above.
E. None of the above.

10. Define *torque*.
Torque is the amount of turning or twisting force applied to a threaded fastener or part.

Name ____________________

D 11. Spanner wrenches are ______.
A. wrenches with drive lugs
B. designed to turn flush type threaded fittings
C. designed to turn recessed type threaded fittings
D. All of the above.
E. None of the above.

12. List the three basic types of spanner wrenches.
Hook, pin, and end.

13. There are many safety precautions that should be observed when using wrenches. List four of them.
Any four of the following: always pull on a wrench; never push; select a wrench that fits properly; never hammer on a wrench to loosen a stubborn fastener; rather than lengthening a wrench handle for additional leverage, use a larger wrench; clean any grease or oil off the handle and the floor in the work area before using a wrench; never try to use a wrench on moving machinery.

14. Identify the screwdriver tips shown below.

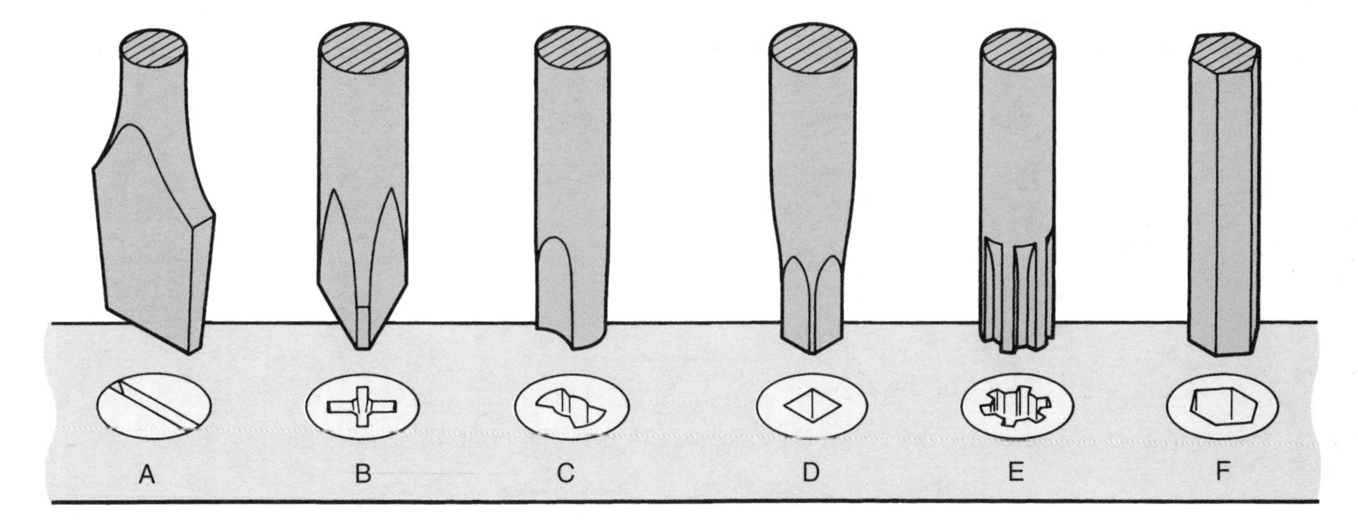

A. Standard
B. Phillips
C. Clutch
D. Square
E. Torx®
F. Hex

15. When are the following chisels used?
A. Flat: For general cutting.
B. Cape: To cut grooves.
C. Round nose: To cut radii and round grooves.
D. Diamond point: For squaring corners.

entire length 16. For long hacksaw blade life, make sure you use the ______ of the blade when cutting.

B 17. Mount the work solidly and as close to the vise as practical for sawing, otherwise ______.

A. the blade will bind and ruin the handle
B. chatter and vibration will dull the teeth
C. the blade may heat up and ruin the teeth
D. All of the above.
E. None of the above.

A 18. Should a blade break or dull before completing the cut, you should *not* continue in the same cut with a new blade because the ______.

A. cut will be too narrow for the new blade
B. cut will be too wide for the new blade
C. You should continue in the same cut with a new blade.
D. All of the above.
E. None of the above.

A 19. The number of teeth per inch of a blade has an important bearing on the shape of the material being cut. For best results, three or more teeth should be cutting at all times; otherwise ______.

A. the teeth will straddle the section being cut and snap off
B. great pressure will be required to make the teeth cut
C. the teeth will clog with chips
D. All of the above.
E. None of the above.

C 20. Files are classified by their shape. File shape is determined by the ______.

A. length and width of the tool
B. taper and thickness
C. general outline and cross section
D. All of the above.
E. None of the above.

21. Files are also classified by the cut of their teeth. Identify the cuts illustrated below.

A. Single cut
B. Double cut
C. Rasp
D. Curved tooth

Name ____

draw 22. When a file is pushed or pulled across the work, it is called _____ filing.

accuracy 23. Reaming is done when a hole must be finished with smoothness and _____.

C 24. A(n) _____ reamer is used when a hole must be cut a few thousandths inch over nominal size for fitting purposes.

A. taper hand
B. spiral-fluted
C. expansion hand
D. All of the above.
E. None of the above.

B 25. A(n) _____ reamer should be used when reaming a hole with a keyway or other interruption.

A. taper hand
B. spiral-fluted
C. expansion hand
D. All of the above.
E. None of the above.

26. List three precautions that should be observed when using hand reamers.

Any three of the following: to prevent injury, remove all burrs from holes; never use your hands to remove chips and cutting fluid from the reamer (use a piece of cotton waste); store reamers carefully so they do not touch one another, they should never be stored loose or thrown into a drawer with other tools; clamp work solidly before starting to ream; do not use compressed air to remove chips and cutting fluid or to clean a reamed hole.

E 27. The drill used prior to tapping is called a _____.

A. special drill
B. threading drill
C. thread cutting drill
D. All of the above.
E. None of the above.

28. Three different taps must be used to thread a blind hole to the bottom. List them in the order they are used.

In order: taper, plug, and bottom.

29. It has been established that a hole to be tapped must be *smaller* than the desired thread diameter. For example: a 5/16″ hole must be drilled for a 3/8-16UNC thread. How much *larger* must a shaft be machined to receive a 3/8-16UNC external thread? Why?
Diameter must be same size as threads.

30. Ragged threads are the most common problem encountered when cutting external threads with a die. They can be caused by:
Any of the following: too little or lack of cutting oil, dull die cutters; stock too large for threads being cut; die not started square; one set of cutters could be upside down when using a two part die.

B 31. An abrasive is _____.
A. a soft substance
B. used to wear away another material
C. used to strengthen metal
D. All of the above.
E. None of the above.

Emery 32. _____ is a natural abrasive.

Silicon carbide 33. _____ is one of the hardest and sharpest of the synthetic abrasives.

34. In the space provided below, record some of the *unsafe* practices you have observed in the shop by people using hand tools. Do *not* use student/trainee names.
Evaluate individually.

CHAPTER

8

Fasteners

Name ______________________________ Date ____________ Class ____________

Learning Objectives

After studying this chapter, you will be able to:

- Identify several types of fasteners.
- Explain why inch-based fasteners are not interchangeable with metric-based fasteners.
- Describe how some fasteners are used.
- Select the proper fastening technique for a specific job.
- Describe adhesive fastening techniques.

Carefully study the chapter, then answer the following questions.

___C___ 1. Threaded fasteners are often used because _____.

A. they are standard sizes

B. they are readily available

C. they permit work to be assembled and disassembled without damage to the parts

D. All of the above.

E. None of the above.

2. List the correct nomenclature of the two thread series shown below. Place your answers in the appropriate spaces.

UNIFIED NATIONAL
COARSE THREAD SERIES

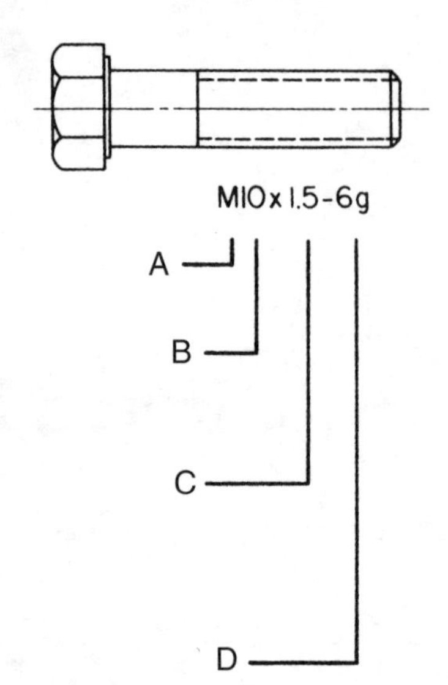

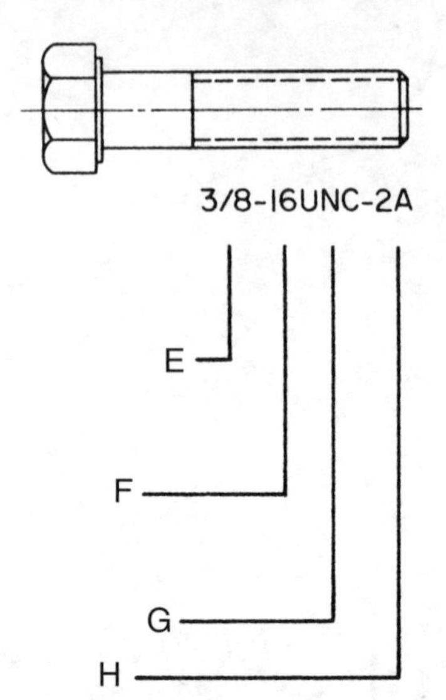

A. Thread symbol for ISO (metric) thread
B. Major diameter of threads in millimeters
C. Pitch of threads in millimeters
D. Thread tolerance class symbol (class of fit)
E. Major diameter of thread in inches
F. Threads per inch (pitch = 1/number of threads per inch)
G. Thread series (Unified National Coarse)
H. Class of fit (thread tolerance)

3. Identify the machine screws illustrated below.

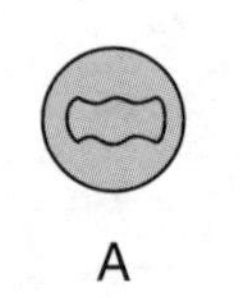

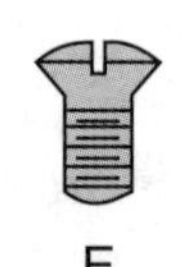

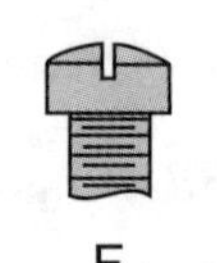

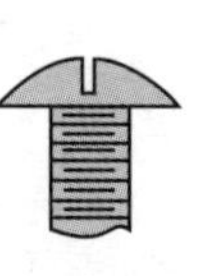

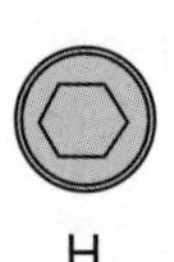

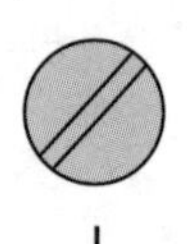

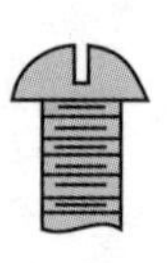

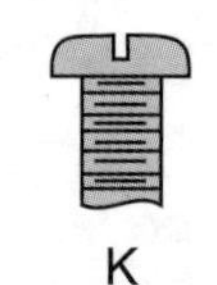

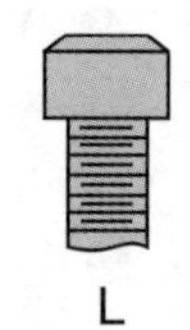

A. Clutch
B. Cross Recess Type 1
C. Cross Recess Type 2
D. Flat
E. Oval
F. Fillister
G. Truss
H. Socket
I. Slotted
J. Round
K. Pan
L. Socket

bolts 4. Machine _____ are employed to assemble parts that do not require close tolerances.

Name ________________________________

Cap 5. _____ screws are found in assemblies that require higher quality fasteners with a more finished appearance.

D 6. Set screws are used to _____.

A. prevent pulleys from slipping on shafts
B. hold collars in place on shafts
C. hold shafts in place on assemblies
D. All of the above.
E. None of the above.

D 7. Nuts for most fasteners have _____ shapes.

A. square
B. hexagonal
C. circular
D. Both A and B.
E. None of the above.

jam 8. The _____ nut is thinner than the standard nut and is frequently used to lock regular nuts in place.

washer 9. A(n) _____ permits a bolt or nut to be tightened without damaging the work surface.

10. Identify the washers illustrated below.

A B C D E

A. Internal-external
B. Internal
C. External
D. Countersunk
E. Split-ring

11. How are drive screws installed?
They are hammered into a drilled or punched hole of the proper size.

12. When are dowel pins employed?
When parts must be accurately positioned and held in absolute relation to one another.

cotter pin 13. The ______ is fitted into a hole drilled crosswise in a shaft to prevent parts from slipping on, or falling off of a shaft.

retaining 14. The grooves needed to install ______ rings eliminate many machining operations.

permanent 15. Rivets are used to make ______ assemblies.

C 16. Blind rivets have been developed for applications where ______.

A. no tool is available to put them in place
B. heavier rivets are required
C. the joint is available from only one side
D. All of the above.
E. None of the above.

Match each word in the left column with the most correct description in the right column.

C 17. Keyseat — A. Head design permits easier removal from assembly.
E 18. Keyway — B. Usually one-fourth the shaft diameter.
B 19. Square key — C. A slot cut into the shaft of a gear or pulley.
F 20. Pratt & Whitney key — D. Semicircular.
A 21. Gib head key — E. A slot cut into the hub of a gear or pulley.
D 22. Woodruff key — F. Rounded at both ends.

23. List the steps, in their proper sequence, involved when adhesives are used to join metal sections.

A. Surface preparation
B. Adhesive preparation
C. Adhesive application
D. Assembly
E. Bond development

24. List at least four (4) safety precautions that should be observed when working with adhesives.

Evaluate individually. Answers may include: Wear approved eye protection, wear disposable plastic gloves when preparing and applying adhesives, keep fingers away from your eyes and mouth, carefully follow the manufacturer's direction for mixing and applying adhesives, mix only the amount of adhesive you will need, do not apply adhesives near open flames.

CHAPTER

9

Jigs and Fixtures

Name ________________________________ Date ____________ Class ____________

Learning Objectives

After studying this chapter, you will be able to:

- Describe a jig.
- Describe a fixture.
- Elaborate on the classifications of jigs and fixtures.
- Explain why jigs and fixtures are used.
- Describe a tombstone.

Carefully study the chapter, then answer the following questions.

manufacturing costs 1. The use of jigs and fixtures helps reduce _____.

B 2. Jigs are _____.

A. always mounted to the drill press table

B. usually nested between guide bars

C. only used when a small number of parts must be produced

D. All of the above.

E. None of the above.

3. Identify the parts indicated in the illustration below.
   A. Jig
   B. Guide bars
   C. Drill press table

D 4. The simplest form of the open jig _____.
   A. is called a plate jig
   B. uses a plate with holes to guide the drill
   C. fits over the work
   D. All of the above.
   E. None of the above.

B 5. A box or closed jig _____ the work.
   A. fits over
   B. encloses
   C. sits parallel to
   D. All of the above.
   E. None of the above.

C 6. The box or closed jig is used when holes must be drilled _____.
   A. only at one end of the piece
   B. all in one direction
   C. in several directions
   D. All of the above.
   E. None of the above.

A 7. Slip bushings are used _____.
   A. to guide the drills
   B. when spot facing
   C. when mounting the jig to the table
   D. All of the above.
   E. None of the above.

Name ____

___A___ 8. Fixtures are used to _____.

A. hold work while machining operations are performed

B. guide a cutting tool

C. mount the jig to the table

D. All of the above.

E. None of the above.

___D___ 9. Jigs and fixtures _____.

A. can be complex in design

B. are designed for specific jobs

C. may have bodies that are welded

D. All of the above.

E. None of the above.

___holding___ 10. Special fixture-_____ devices have been developed for machining centers and other CNC machine tools.

Notes

CHAPTER 10

Cutting Fluids

Name ______________________ Date ____________ Class ____________

Learning Objectives

After studying this chapter, you will be able to:

- Understand why cutting fluids are necessary.
- List the types of cutting fluids.
- Describe each type of cutting fluid.
- Discuss how cutting fluids should be applied.

Carefully study the chapter, then answer the following questions.

mineral 1. Cutting oils made from _____ oil may be used straight or combined with additives.

2. Mineral oils are best suited for:
light-duty (low speed, light feed) operations where high levels of cooling and lubrication are not required

D 3. Mineral oils _____.
A. are often combined with vegetable oils
B. are noncorrosive
C. can contain sulfur
D. All of the above.
E. None of the above.

4. What precautions must be taken in situations where coolant mists or vapors are present?
An approved respirator must be worn.

Bacteriostats 5. _____ are added to some cutting fluids to control or regulate the growth of bacteria.

D 6. Emulsifiable oils, also known as soluble oils, _____.

A. provide increased cooling capacity in some applications
B. range from milky to transparent in appearance
C. are composed of oil droplets
D. All of the above.
E. None of the above.

water-based 7. When machining magnesium, _____ cutting fluids must *never* be used.

wetting 8. A _____ agent is often added to chemical cutting fluids to provide moderate lubricating qualities.

9. List three disadvantages of using chemical and semichemical cutting fluids.
Some formulas have minimal lubricating qualities; they may cause skin irritation in some workers; and when they become contaminated with other oils, disposal can be problematic.

D 10. Compressed air _____.

A. is the most commonly used gaseous coolant
B. cools by forced convection
C. can be dangerous
D. All of the above.
E. None of the above.

chips 11. When using cutting fluids, the coolant nozzles must be positioned carefully so that, in addition to cooling the work area, the cutting fluid will also aid in removing _____.

cooling rates 12. Since carbide tooling operates at higher cutting speeds and generates higher cutting temperatures, cutting fluids that have high _____ should be used in such applications.

CHAPTER 11

Sawing and Cutoff Machines

Name ______________________ Date __________ Class __________

Learning Objectives

After studying this chapter, you will be able to:

- Identify the various types of sawing and cutoff machines.
- Select the correct machine for the job to be done.
- Mount a blade and prepare the machine for use.
- Position the work for the most efficient cutting.
- Safely operate sawing and cutoff machines.

Carefully study the chapter, then answer the following questions.

___C___ 1. The _____ feed of the reciprocating power hacksaw, yields a pressure on the blade that is uniform regardless of the number of teeth in contact with the work.

A. positive
B. negative pressure
C. definite pressure
D. All of the above.
E. None of the above.

___B___ 2. Large sections and soft materials require the use of _____.

A. light feed pressure
B. coarse-tooth blades
C. fine tooth blades
D. All of the above.
E. None of the above.

C 3. _____ blades should be used where safety requirements demand a shatterproof blade and for cutting odd-shaped work if there is a possibility of the work coming loose in the vise.

A. All-hard
B. Standard tooth
C. Flexible-back
D. All of the above.
E. None of the above.

4. Identify the set of band saw blades shown below.
A. Raker
B. Wavy

A B

5. What does a regular band machine maintenance program typically include?
Checking wheel alignment, guide alignment, feed pressure, hydraulic systems.

6. Most sawing problems can be corrected or prevented. What are the causes of the following problems?
A. Broken blade: Blade dropped on work. Loose blade or excessive speed.
B. Crooked cut: Usually caused by worn blade.
C. Blade pin holes breaking out: Dirty mounting plates or too much tension on blade.
D. Premature blade teeth wear: Insufficient pressure/excessive pressure. Lack of coolant or poorly adjusted machine.
E. Teeth strip off: Starting cut on sharp corner. Less than three teeth cutting, blade too fine or too coarse.

7. What are the two classifications of the abrasive cutoff saw?
Wet, Dry

accurate 8. The toothed blade on the cold circular saw is capable of producing very _____ cuts.

teeth 9. A friction saw blade may or may not have _____.

Name ______________________________

billets 10. The friction saw finds many applications in steel mills to cut ______.

11. What do *you* believe are the three most dangerous aspects of operating a power saw?
Evaluate individually. Answers may include: Too much tension on a saw blade can shatter the blade. Band saw blades are long and springy and can uncoil suddenly. Avoid standing directly in line with the blade when operating a circular cutoff saw.

Notes

CHAPTER 12

Drills and Drilling Machines

Name ______________________ Date __________ Class __________

Learning Objectives

After studying this chapter, you will be able to:

- Select and safely use the correct drills and drilling machine for a given job.
- Explain the safety rules that pertain to drilling operations.
- Identify and describe common drills and drill-holding devices.
- Make safe setups on a drill press.
- Sharpen a twist drill.
- Describe basic drilling operations.

Carefully study the chapter, then answer the following questions.

__drill press__ 1. The _____ is the most common drilling machine.

__B__ 2. The size of a drill press is determined by _____.

A. the size of its motor
B. the largest diameter circular piece that can be drilled on center
C. the largest size tool shank that can be used
D. All of the above.
E. None of the above.

__C__ 3. _____ can be used to drill holes in small workpieces but do not have as many capabilities as the floor models.

A. Radial drill presses
B. Mini drill presses
C. Bench drill presses
D. All of the above.
E. None of the above.

A 4. _____ are designed to handle very large drilling work.
A. Radial drill presses
B. Floor model drill presses
C. Electric hand drills
D. All of the above.
E. None of the above.

D 5. Portable magnetic drills can be positioned in a(n) _____ position when drilling.
A. vertical
B. horizontal
C. upright
D. All of the above.
E. None of the above.

D 6. Drilling machines are used for _____.
A. tapping
B. countersinking
C. cutting round holes
D. All of the above.
E. None of the above.

7. Drills made from _____ can be operated at much higher cutting speeds than drills made from carbon steel.
high-speed steel (HSS)

8. Coating drills with _____ greatly increases tool life.
titanium nitride

B 9. _____ refers to the sharp edge of the extreme tip of the drill.
A. The heel
B. Dead center
C. Lip clearance
D. All of the above.
E. None of the above.

A 10. A tang assists in driving the tool, and _____.
A. provides a means of separating the taper from the holding device
B. mounts into the chuck
C. determines lip clearance
D. All of the above.
E. None of the above.

Name ____

___D___ 11. The body is the portion of the drill between the point and the shank. It has _____.

A. a metal column that separates the flutes

B. two or more spiral grooves that run along its length

C. a narrow strip extending along its entire length

D. All of the above.

E. None of the above.

___B___ 12. A(n) _____ has coolant holes through the body.

A. three-flute core drill

B. coolant-hole drill

C. step drill

D. All of the above.

E. None of the above.

___E___ 13. _____, which use low-cost carbide inserts, are capable of drilling at much higher speeds than high-speed steel drills.

A. Conventional twist drills

B. Three-flute core drills

C. Straight-flute gun drills

D. All of the above.

E. None of the above.

___drill margins___ 14. When a micrometer is used for measuring drills, the measurement is made across the _____ unless the drill is worn.

___D___ 15. A drill is held in the drill press with a _____.

A. tapered spindle

B. chuck

C. web

D. Either A or B.

E. None of the above.

16. Identify the drilling machine accessories shown below.

A

B

C

A. Sleeve

B. Socket

C. Drift

17. What is the drilling machine accessory "A" (above) used for?
To enlarge smaller taper shanks to fit the drill press spindle.

18. What is the drilling machine accessory "B" (above) used for?
To reduce a taper shank so it will fit the drill press spindle.

19. What is the drilling machine accessory "C" (above) used for?
To separate taper shanks from sleeves, sockets, or drills from drill press spindles.

Parallels 20. _____ are often used to level the work and raise it above the vise base.

angular 21. A(n) _____ vise permits angular drilling without tilting the drill press table.

D 22. When setting work up on a drill press table, _____.
A. place T-bolts as close to the work as possible
B. place a washer between the nut and the holding device
C. fit the T-bolts into the drill press table
D. All of the above.
E. None of the above.

U-strap 23. When the work-holding device must bridge the work, a(n) _____ clamp should be used because it can straddle the drill and not interfere with the drilling operation.

A 24. Drill cutting speed, also known as peripheral speed, refers to the _____.
A. distance that the drill cutting edge circumference travels per minute
B. revolutions per minute (rpm) of the drill
C. distance the drill is moved into the work with each revolution
D. All of the above.
E. None of the above.

25. The spiral shape of the drill flute _____ (causes/does not cause) the drill to pull itself into the work.
does not cause

Name ____

Using the formula below, calculate problems 26–29. Round your answer off to the nearest whole number.

Formula: $\text{rpm} = \frac{4 \times CS}{D}$

26. At what speed (rpm) must a 1/2″ diameter high-speed drill rotate to drill aluminum? Recommended cutting speed is 300 fpm. Use the space below for your calculations.
    2400 rpm

27. At what speed (rpm) must a 1/4″ diameter high-speed drill rotate to drill free machining brass? Recommended cutting speed is 250 fpm. Use the space below for your calculations.
    4000 rpm

28. At what speed (rpm) must a 5/8″ diameter high-speed drill rotate to drill malleable iron? Recommended cutting speed is 75 fpm. Use the space below for your calculations.
    480 rpm

29. At what speed (rpm) must a 9/16″ diameter high-speed steel drill rotate to drill free machining steel? Recommended cutting speed is 90 fpm. Use the space below for your calculations.
    640 rpm

30. How can you be sure a drill point is sharpened at the proper angle?
    Check with drill point gage.

31. After sharpening, how can drills be tested?
    By drilling a hole in soft metal and observing chip formation

32. How can you tell if a drill has been properly sharpened?
When properly sharpened, chips will come out of the flutes in curled spirals of equal size and length.

D 33. Even when work is properly centered, a large drill may "drift" when starting a hole due to _____.
A. an improperly sharpened drill
B. hard spots in the metal
C. the fact that large drills have a tendency to do so
D. All of the above.
E. None of the above.

V-blocks 34. Since drilling in the curved surface of round stock is more difficult, it is advisable to use _____ to hold the round material in a vise or clamped directly to the table.

35. Why is a hole countersunk?
To receive flat-headed fasteners.

36. Why is a hole spotfaced?
The operation machines a flat circular area on a rough surface to provide a bearing surface for the head of a bolt, washer, or nut.

B 37. A _____ reamer is designed to cut on its end.
A. jobber's
B. rose chucking
C. machine
D. All of the above.
E. None of the above.

shell 38. A(n) _____ reamer is a reamer that is mounted on a special arbor that can be used with several reamer sizes.

D 39. In a microdrilling operation, _____.
A. the drill is repeatedly inserted and removed from the hole
B. a pecking technique is used to cut small diameter holes
C. pecking is necessary for chip removal
D. All of the above.
E. None of the above.

Name ____

Additional Activities

40. Prepare a test piece to the specifications shown below. You will be graded on the accuracy of hole diameter and hole spacing.
    Evaluate individually.

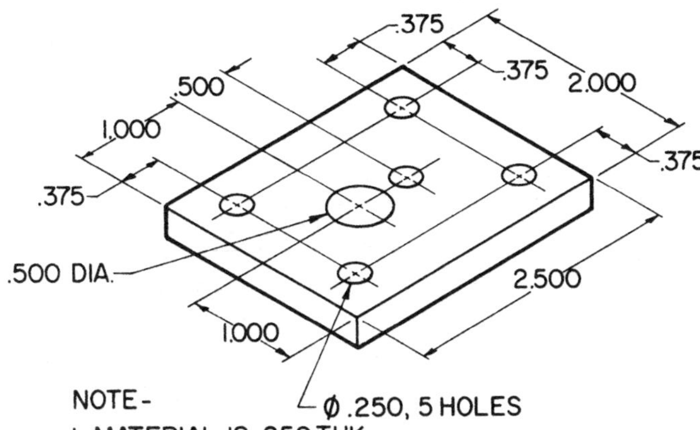

41. Sharpen a drill that has become dull. After sharpening, test it by drilling a hole and checking its size against the drill's diameter.
    Evaluate individually.

    A. Drill size: ______________________________

    B. Hole size: ______________________________

Additional activities 40–41 involve instructor's permission and/or group participation. Review activities and assign as applicable or as time permits.

Notes

CHAPTER 13

Offhand Grinding

Name ______________________ Date ________ Class ________

Learning Objectives

After studying this chapter, you will be able to:

- Identify the various types of offhand grinders.
- Dress and true a grinding wheel.
- Prepare a grinder for safe operation.
- Use an offhand grinder safely.
- List safety rules for offhand grinding.

Carefully study the chapter, then answer the following questions.

1. _____ are heavy-duty versions of the belt and disc sanders found in woodworking.
   Abrasive belt grinders

D 2. In offhand grinding operations, _____.
   A. great accuracy is *not* required
   B. grinding is done on a bench, pedestal, or belt grinder
   C. the part is held in the operator's hands
   D. All of the above.
   E. None of the above.

roughing 3. The bench grinder usually has two wheels. One wheel is coarse for _____.

finish 4. The bench grinder usually has two wheels. One wheel is fine for _____ grinding.

C 5. A wet-type pedestal grinder _____.
A. has no provisions for cooling the work
B. requires that the part be dipped in water
C. keeps the wheels constantly flooded with fluid
D. All of the above.
E. None of the above.

D 6. For safety reasons, the tool rest on a bench or pedestal grinder should be a maximum of _____.
A. 1/16″
B. 1.5 mm
C. 0.15 cm
D. All of the above.
E. None of the above.

D 7. Grinding wheels should be examined frequently for _____.
A. roundness
B. concentricity
C. soundness
D. All of the above.
E. None of the above.

wheel dresser 8. A(n) _____ is used to true the wheel and remove any glaze that may have formed during grinding operations.

C 9. To obtain maximum efficiency from a grinder, you should _____.
A. hold the work in one place on the wheel face
B. grind using the sides of the wheel
C. use the face of the wheel, *not* the sides
D. All of the above.
E. None of the above.

10. Moving the work across the wheel face can prevent the formation of _____.
grooves/ridges

greater 11. Avoid operating a grinding wheel at speeds _____ than that recommended by the wheel manufacturer.

12. Why shouldn't a piece of cloth be used to hold work while it is being ground?
Serious injury could result if the cloth is pulled into the wheel.

Name ______________________________

13. List four more safety precautions that should be observed when operating offhand grinders.
Evaluate individually. Answers may include: Wear goggles or a face shield when performing grinding operations, be sure all wheel guards and safety devices are in place before attempting to use a grinder, stand to one side of the machine during operation, keep your hands clear of the rotating wheel.

__A__ 14. _____ wheels should be used to grind the carbide tip on carbide tools.
A. Diamond-impregnated
B. Titanium
C. Aluminum oxide
D. All of the above.
E. None of the above.

__C__ 15. The face of the wet grinder is crowned slightly to minimize the possibility of a carbide tip lathe cutting tool from _____.
A. being ground away too rapidly
B. damaging the wheel by causing excessive wear
C. being damaged or destroyed by excessive heat buildup
D. All of the above.
E. None of the above.

__D__ 16. Portable hand grinders can be _____.
A. powered by electricity or air
B. used for light deburring
C. used to polish dies
D. All of the above.
E. None of the above.

Notes

CHAPTER 14

The Lathe

Name ______________________ Date __________ Class __________

Learning Objectives

After studying this chapter, you will be able to:

- Describe how a lathe operates.
- Identify the various parts of a lathe.
- Safely set up and operate a lathe using various work-holding devices.
- Calculate correct cutting speeds and feeds for lathe operations.
- Perform basic machining operations on a lathe.
- Sharpen lathe cutting tools.

Carefully study the chapter, then answer the following questions.

__B__ 1. The back gear on a lathe system _____.

A. should be engaged while the spindle is rotating
B. provides slower speeds with greater power
C. moves the drive belt to another pulley ratio
D. All of the above.
E. None of the above.

__D__ 2. The spindle on a lathe is hollow and tapered internally to _____.

A. allow the use of a knockout bar
B. receive tools and attachments with taper shanks
C. permit long stock to be turned without dangerous overhang
D. All of the above.
E. None of the above.

C 3. A _____ is seldom used on modern lathes.

A. long taper key spindle
B. cam-lock spindle nose
C. threaded spindle nose
D. All of the above.
E. None of the above.

D 4. The tailstock _____.

A. is essential for drilling operations on a lathe.
B. permits taper turning to be done
C. is necessary to support long work being turned
D. All of the above.
E. None of the above.

index plate 5. The _____ provides instructions on how to set the lathe shift levers for various thread cutting and feed combinations.

lead screw 6. The _____ transmits power to the carriage through a gearing and clutch arrangement in the carriage apron.

7. Identify the parts indicated on the lathe illustrated below.

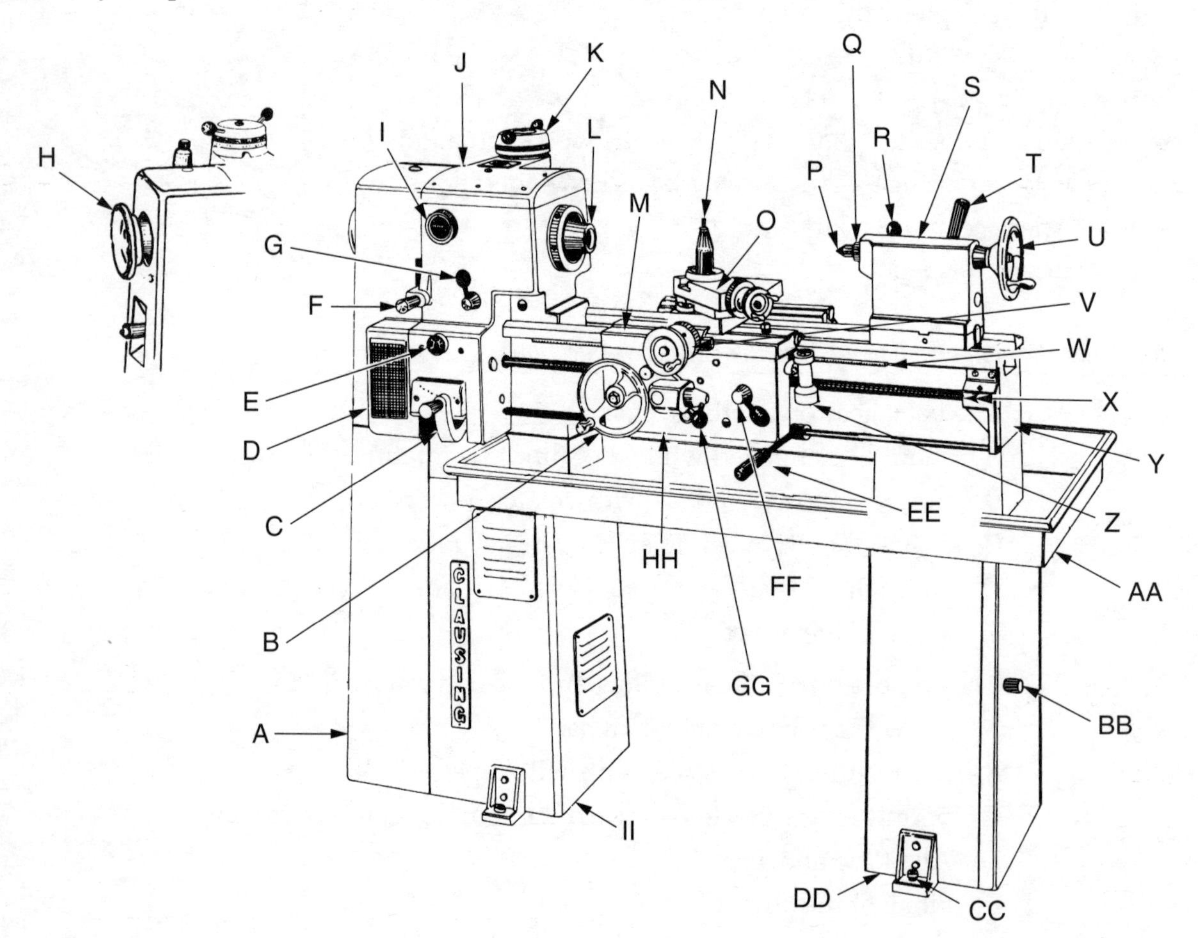

Name ______________________________

A. Motor and gear train cover
B. Carriage handwheel
C. Thread and feed selector lever
D. Quick-change gearbox
E. Selector knob
F. Lead screw direction lever
G. Motor control lever
H. Backgear handwheel
I. Backgear control knob
J. Headstock
K. Variable speed control
L. Spindle
M. Carriage saddle
N. Tool post
O. Compound rest
P. Dead center
Q. Tailstock ram
R. Ram lock
S. Tailstock
T. Tailstock lock lever
U. Handwheel
V. Cross-slide handwheel
W. Rack
X. Lead screw
Y. Bed
Z. Threading dial
AA. Chip pan
BB. Storage compartment door
CC. Leveling screw
DD. Tailstock pedestal
EE. Clutch and brake handle
FF. Half-nut lever
GG. Power feed lever
HH. Carriage apron
II. Headstock pedestal

Use the illustration in question 7 to answer questions 8–15.

D 8. If item D is changed, the cutting tool will _____. (Assume that power is being transmitted through the gear box.)

A. cut deeper
B. move faster or slower
C. cut better
D. move faster or slower if the carriage is engaged to the lead screw
E. None of the above.

D 9. Item K _____.

A. reduces or increases motor speed
B. increases power to the spindle
C. puts tension on the belt
D. changes spindle speed
E. None of the above.

E 10. Item X transmits power from the quick change gear box to the _____.
- A. tailstock
- B. headstock
- C. spindle
- D. back gears
- E. None of the above.

E 11. Item L _____.
- A. is removed with a hammer
- B. supports the work
- C. is lubricated each day
- D. makes the centers line up
- E. None of the above.

B 12. Item FF _____.
- A. causes the cutter bit to move in and out
- B. engages the half-nuts for threading
- C. engages the clutch for automatic power feed
- D. locks the unit to the ways
- E. None of the above.

B 13. Item GG _____.
- A. locks the unit to the ways
- B. engages the clutch for automatic power feed
- C. engages the half-nuts for threading
- D. causes the cutter bit to move up and down
- E. None of the above.

A 14. Item B _____.
- A. moves the entire unit right and left on the ways
- B. moves the cutter bit in and out
- C. engages the unit for threading
- D. locks the unit to the ways
- E. None of the above.

C 15. Item GG engages the _____.
- A. automatic power feed
- B. half-nuts for threading
- C. automatic power cross-feed
- D. unit to the ways
- E. None of the above.

16. A(n) _____ and countersink is usually used to drill center holes.
combination drill

17. How will the work be affected if the headstock center does not run true when working between centers?
Eccentric diameters will result if the headstock center does not run true.

Name ____

___C___ 18. The _____ dog has the setscrew recessed.
A. clamp-type
B. bent-tail standard
C. bent-tail safety
D. All of the above.
E. None of the above.

___B___ 19. The _____ dog has the setscrew exposed.
A. clamp-type
B. bent-tail standard
C. bent-tail safety
D. All of the above.
E. None of the above.

___A___ 20. The _____ dog is used for turning square or rectangular work.
A. clamp-type
B. bent-tail standard
C. bent-tail safety
D. All of the above.
E. None of the above.

___A___ 21. A _____ chuck automatically centers work. All of the jaws move simultaneously.
A. 3-jaw universal
B. Jacobs
C. 4-jaw independent
D. collet
E. None of the above.

___C___ 22. The jaws on a _____ chuck can hold irregular shaped work as each jaw has individual movement.
A. 3-jaw universal
B. Jacobs
C. 4-jaw independent
D. collet
E. None of the above.

___B___ 23. A _____ chuck is normally fitted in the tailstock but can also be used to hold small diameter work for turning if fitted into the headstock.
A. 3-jaw universal
B. Jacobs
C. 4-jaw universal
D. collet
E. None of the above.

collet 24. The chief advantage of a(n) _____ chuck is its ability to center work automatically and maintain accuracy over long periods of hard usage.

25. Why do collet chucks have the disadvantage of being expensive?
A separate collet is required for each different size or shape of stock.

26. What is the most accurate method for centering round stock in a 4-jaw chuck?
Using a dial indicator.

C 27. The jaws of a _____ chuck can be reversed to hold large diameter work. They cannot be reversed on a _____ chuck.
A. 3-jaw independent, 3-jaw universal
B. 3-jaw independent, 4-jaw universal
C. 4-jaw independent, 3-jaw universal
D. None of the above.

28. What is the most important safety precaution to remember when using a chuck?
Be sure to remove the chuck key before turning on the machine.

B 29. When determining whether or not a lathe toolholder is a right- or a left-hand model, you can hold the head of the tool in your hand and note the direction the shank points. The shank of the left-hand toolholder points _____.
A. straight
B. to the left
C. to the right
D. All of the above.

D 30. Cutter bits are ground to cut _____.
A. to the left only
B. to the right only
C. in either direction
D. It depends on the work being done.

D 31. The deep cuts made to remove a large amount of material from a workpiece are called _____.
A. side relief cuts
B. chafing cuts
C. turret cuts
D. None of the above.

round nose 32. A(n) _____ tool is ground flat on the face and designed for lighter turning.

33. What will occur if tools designed for machining steel are *not* honed?
The irregular edge produced by grinding will crumble when used.

Name ______________________________

Chipbreakers 34. _____ are used to break the long continuous chips that are created when machining some metals.

For problems 35–37, use the formula below to calculate the correct rpm for machining the materials given. Round off your answers to the nearest zero. Perform your calculations in the space provided.

$$\text{rpm} = \frac{\text{CS} \times 4}{\text{D}}$$

rpm = revolutions per minute

CS = Cutting speed of the particular metal being machined in feet per minute (fpm)

D = Diameter of work in inches

685 rpm 35. Aluminum, 3 1/2″ diameter, CS = 600 fpm

320 rpm 36. Mild steel, 1.250″ diameter, CS = 100 fpm

84 rpm 37. Tool steel, 2.375″ diameter, CS = 50 fpm

For problems 38–39, use the formula below to calculate the correct rpm for machining the materials given. Cutting speeds may also be given in meters per minute (mpm) when the work diameter is given in millimeters (mm). To find rpm for a given cutting speed in mpm, the meters must be converted to millimeters. This can be accomplished by multiplying the cutting speed by 1000. This conversion is included in the formula below. Perform your calculations in the space provided.

$$\text{rpm} = \frac{\text{CS} \times 1000}{3.14 \times \text{D (mm)}}$$

rpm = revolutions per minute

CS = Cutting speed of the particular metal being machined in meters per minute (mpm)

D = Diameter of work in mm

730 rpm 38. Aluminum, 87 mm diameter, CS = 200 mpm

186 rpm 39. Tool steel, 60 mm diameter, CS = 35 mpm

2″ paintbrush 40. When removing chips from a lathe, it is recommended that you use a(n) _____.

machine oil 41. To prevent rust from forming, a light coating of _____ should be applied to all machined surfaces.

42. What should be done to prevent *springing* when machining long work?
Long work should be center drilled and supported with a tailstock center.

B 43. Facing cuts can be made _____.
A. in an inward direction only
B. in either direction
C. in an outward direction only
D. from the center and fed out only
E. None of the above.

Name ________________________________

44. What does a *rounded nubbin* indicate?
That the cutter is slightly above center.

45. What does a *square-shoulder nubbin* indicate?
That the cutter is below center.

46. *Never* attempt to perform a(n) _____ operation on work being turned between centers.
parting or cutoff

Notes

CHAPTER

15

Other Lathe Operations

Name ______________________ Date __________ Class __________

Learning Objectives

After studying this chapter, you will be able to:

- Perform boring and knurling operations on a lathe.
- Describe how drilling, reaming, filing, polishing, grinding, and milling operations can be performed on a lathe.
- Properly set up steady and follower rests.
- Safely set up and operate a lathe using various work-holding devices.
- Demonstrate familiarity with industrial applications of the lathe.

Carefully study the chapter, then answer the following questions.

1. Why is boring done?
   Boring is employed to enlarge a hole to a specified size where a drill or reamer will not do the job.

reversed 2. When boring on a lathe, the movement of the cross slide screw must be ______.

3. Why must additional front clearance be ground on a boring tool cutter?
   To prevent the bottom of the tool from rubbing on the bored surface.

tool holder 4. A boring bar should extend from the ______ only far enough to do the job.

B ______ 5. A boring bar is positioned _____ for machining.

A. above center

B. on center

C. below center

D. All of the above.

E. None of the above.

6. Because of the slender nature of some boring bars, *chatter* is more likely to occur when boring than when doing external machining. List five ways in which this can usually be eliminated.

Any order: Using a slower spindle speed, reducing tool overhang, grinding a smaller radius on the nose of the cutting tool, placing a weight on the back overhang of the boring bar, placing the tool slightly below center.

taper ______ 7. When drilling on a lathe, drills larger than 1/2″ (12.5 mm) in diameter usually have _____ shanks.

E ______ 8. When drilling with drills 1/2″ (12.5 mm) in diameter or smaller, a starting point made with a(n) _____ is adequate.

A. combination drill

B. spade drill

C. countersink (center drill)

D. Both A and B.

E. Both A and C.

dead center ______ 9. Holes over 1/2″ (12.5 mm) in diameter require a pilot hole, which should have a diameter equal to the width of the larger drill's _____.

10. What precaution must be observed when through drilling a hole in work held in a chuck?

Check for adequate clearance between the back of the work and the chuck face.

When reaming, the hole is drilled slightly undersize. The allowance for reaming depends upon hole size. For questions 11–15, provide the proper allowance in inches and millimeters.

11. With a hole size above 1.5″ (37.5 mm) in diameter, allow _____ (_____).

0.030″ (0.8 mm)

12. With a hole size up to 1/4″ (6.5 mm) in diameter, allow _____ (_____).

0.010″ (0.25 mm)

13. With a hole size from 1.0″ (25.0 mm) to 1.5″ (37.5 mm) in diameter, allow _____ (_____).

0.025″ (0.6 mm)

Name ____

14. With a hole size from 1/4″ (6.5 mm) to 1/2″ (12.5 mm) in diameter, allow _____ (_____).
    0.015″ (0.4 mm)

15. With a hole size from 1/2″ (12.5 mm) to 1.0″ (25.0 mm) in diameter, allow _____ (_____).
    0.020″ (0.5 mm)

16. What is a common cause of a *double-cut knurl*?
    One wheel is dull.

17. List three precautions that should be observed when filing on a lathe.
    Move the carriage out of the way, use the left-hand method of filing, and keep the file moving.

18. List three uses of steady and follower rests.
    Evaluate individually. Refer to Section 15.5. Answers may include: Provide additional support for long and thin workpieces, keep the workpiece from springing or bending away from the cutting tool, and reduce chattering.

centers 19. Work mounted on a mandrel is usually machined between _____.

20. Name three uses for the *lathe tool post grinder*.
    Any order: Grind shafts, true lathe centers, and sharpen reamers and milling cutters.

21. How should a lathe be prepared for grinding operations?
    Evaluate individually. Refer to Section 15.7.1. Answers may include: The lathe bed, cross slide, and other parts should be covered with canvas or heavy kraft paper. Place a small tray of water or oil below the grinding wheel.

22. What does the term *spark out* mean?
    The point at which the grinding wheel no longer cuts.

A 23. When performing internal grinding on the lathe, the work and the grinding wheel *must* rotate _____.

A. in opposite directions
B. in the same direction
C. at different rates of speed
D. Both A and C.
E. Both B and C.

turret 24. Limited production runs are sometimes produced on a manually operated _____ lathe.

Swiss-type 25. A(n) _____ machine can produce tiny precision parts in quantity.

26. A(n) _____ is used for work that is too heavy or too large to be turned in a horizontal position.
vertical lathe or boring mill

CHAPTER 16 Cutting Tapers and Screw Threads on the Lathe

Name ______________________ Date __________ Class __________

Learning Objectives

After studying this chapter, you will be able to:

- Describe how a taper is turned on a lathe.
- Calculate tailstock setover for turning a taper.
- Safely set up and operate a lathe for taper turning.
- Describe the various forms of screw threads.
- Cut screw threads on a lathe.

Carefully study the chapter, then answer the following questions.

D 1. Taper can be stated _____.
  A. in taper per inch or foot
  B. as degrees or as a ratio
  C. mm per 25 mm
  D. All of the above.
  E. None of the above.

2. Only external tapers can be machined with the _____ method.
offset tailstock or tailstock setover

Machine adjustments must be calculated for each tapering unit. The information that follows will enable you to calculate the necessary adjustment (tailstock setover) for problems 3–5.

Formulas: When taper per inch is known: $\text{Offset} = \dfrac{L \times TPI}{2}$

When taper per foot is known: $\text{Offset} = \dfrac{L \times TPF}{24}$

When dimensions of tapered section are known but TPI or TPF is *not* given: $\text{Offset} = \dfrac{L \times (D - d)}{2 \times l}$

Where:

TPI = Taper per inch

TPF = Taper per foot

D = Diameter at large end

d = Diameter at small end

l = Length of taper

L = Total length of piece

0.025″ ____________

3. Compute the tailstock setover (offset) for the following job. Show your work.
   Taper per inch = 0.125
   Total length of piece = 4.00″

0.044″ ____________

4. Compute the tailstock setover for the following job. Show your work.
   Taper per foot = 0.125
   Total length of piece = 8.500′

1.50″ ____________

5. Compute the tailstock setover for the following job. Show your work.
   Large diameter = 2.500
   Small diameter = 1.500
   Length of taper = 3.000
   Length of piece = 9.000

Name ______________________________

The common tapers used to hold cutting tools and tool holders will not change with the metric system. Usually, these tapers are given in inches per foot or inches per inch or as a relationship. If the taper is given in inches per inch, then it will be given in millimeters per 25 millimeters. If the taper is given as a ratio, this will not change. Answer problems 6–8 using the information below.

Other-than-standard tapers can be shown in several different ways:

Taper per millimeter (T/mm). For example, a taper of 0.002 mm per mm is expressed as TAPER 0.002:1.

Dimensions shown in mm where the taper is cut on the total length of the work.

Formulas: When taper per mm is known:

$$\text{Offset} = \frac{L/mm \times T/mm}{2}$$

When dimensions of tapered section are known but T/mm is not given:

$$\text{Offset} = \frac{L/mm \times (D/mm - d/mm)}{2 \times l/mm}$$

Where:

T/mm = Taper per millimeter
D/mm = Diameter at large end in millimeters
d/mm = Diameter at small end in millimeters
l/mm = Length of taper in millimeters
L/mm = Total length of piece in millimeters

0.876 mm ______________

6. Compute the tailstock setover for the following job. Show your work.
   Taper per mm = TAPER 0.002:1
   L/mm = 876 mm

175.0 mm ______________

7. Compute the tailstock setover for the following job. Show your work.
   D/mm = 150.0
   d/mm = 100.0
   L/mm = 875.0
   l/mm = 125.0

14 mm ______________

8. Compute the tailstock setover for the following job. Show your work.
   D/mm = 225.0
   d/mm = 125.0
   L/mm = 1000.0
   l/mm = 875.0

D 9. When an ample tolerance is allowed (±0.015″ or 0.05 mm), the tailstock setover can be made by using a rule and measuring the distance between ______.
   A. center points
   B. witness lines at base of tailstock
   C. the headstock and tailstock
   D. Both A and B.
   E. All of the above.

10. An accurate setover can be made by using the ______ on the lathe cross slide or a dial indicator.
micrometer collar

11. Why should ball-tipped centers be used when cutting tapers by the tailstock setover method?
Lessens pressure on the tail center.

12. There are two types of taper attachments. Identify and briefly describe each.
Plain taper attachment: Requires the cross-slide screw to be disengaged from the cross-slide feed nut. The cutting tool must be advanced by the compound rest feed screw.
Taper attachment: It is not necessary to disengage the cross-slide feed nut.

13. What is the disadvantage of turning a taper with a square-nose tool?
Can only cut short tapers.

14. List the two basic methods of testing the accuracy of machined tapers.
Measuring tapers by comparison using plug and ring gages, or direct measurement of the taper using a taper test gage, gage blocks and parallels, or a sine bar.

15. What is the pitch of inch-based threads?
One divided by the number of threads per inch or 1/N (N = number of threads per inch).

16. When cutting V threads, the ______ is used for grinding the cutting tool and positioning it to cut the threads.
center gage or fishtail

Name ______________________________

17. What two purposes does the thread end groove serve?
Provides a place to stop the threading tool at the end of its cut and permits the nut to be run up to the end of the thread.

C 18. When cutting threads on the lathe, a _____ may be used since the tool must be removed from the work after each cut and repositioned before the next cut can be started.
    A. thread end groove
    B. center gage
    C. thread cutting stop
    D. All of the above.
    E. None of the above.

19. What is a thread dial?
A device on the lathe that indicates when to engage the half-nuts to permit the tool to follow exactly in the original cut.

not engaged 20. The face of the thread dial rotates when the half-nuts are _____ (engaged/not engaged).

D 21. Always check the thread pitch after the first light cut with a _____.
    A. half-nut
    B. rule
    C. screw pitch gage
    D. Both B and C.
    E. None of the above.

22. After replacing a broken cutting tool, how do you realign it with the portion of the thread already cut?
Evaluate individually. Refer to Section 14.6.4. Answers may include: Set the tool on center and position it with a center gage. Engage the half-nuts at the proper thread dial graduation. Move the tool back from the work and rotate the spindle until the tool reaches a position about halfway down the threaded section. Using the compound rest screw and the cross-slide screw, align the tool in the existing thread. Reset the thread cutting stop after the tool has been aligned.

For problems 23–37, using the 3-wire method for measuring screw threads, calculate the correct measurement over the wire for the threads given. Use the wire size given.

Formulas: $M = D + 3G - \frac{1.5155}{N}$

Where:
M = Measurement over the N wires
D = Major diameter of thread
G = Diameter of wires
N = Number of threads per inch

0.324″ ______ 23. 5/16-18UNC (Wire size 0.032″)

0.384″ ______ 24. 3/8-24UNF (Wire size 0.024″)

0.457″ ______ 25. 7/16-20UNF (Wire size 0.032″)

0.515″ ______ 26. 1/2-13 UNC (Wire size 0.044″)

0.763″ ______ 27. 3/4-16UNF (Wire size 0.036″)

0.899″ ______ 28. 7/8-9UNC (Wire size 0.064″)

1.154″ ______ 29. 1 1/8-7UNC (Wire size 0.082″)

Name ______________________________

0.257″ 30. 1/4-32UNS (Wire size 0.018″)

0.225″ 31. #12-24UNC (0.216″ diameter) (Wire size 0.0240″)

0.153″ 32. 7/16-14UNC (Wire size 0.0412″)

0.580″ 33. 9/16-12UNC (Wire size 0.0481″)

0.645″ 34. 5/8-11UNC (Wire size 0.0525″)

0.771″ 35. 3/4-10UNC (Wire size 0.0577″)

1.281″ 36. 1 1/4-7UNC (Wire size 0.0825″)

1.411″ 37. 1 3/8-6UNC (Wire size 00.0962″)

D 38. Cutting left-hand threads requires _____.
A. that the carriage travels toward the tailstock
B. changing the lead screw rotation
C. pivoting the compound to the left
D. All of the above.
E. None of the above.

B 39. When cutting internal threads, tool infeed and removal from the cut are _____ when cutting external threads.
A. the same as those used
B. the reverse of those used
C. not necessary
D. All of the above.
E. None of the above.

B 40. When cutting tapered threads to obtain a fluid- or gas-tight joint, the threading tool must be positioned _____.
A. in relation to the taper itself
B. in relation to the centerline of the taper
C. in relation to the setover
D. Either A or B.
E. None of the above.

CHAPTER

17

Broaching Operations

Name ________________ Date ________ Class ________

Learning Objectives

After studying this chapter, you will be able to:

- Describe the broaching operation.
- Explain the advantages of broaching.
- Set up and cut a keyway using a keyway broach and an arbor press.

Carefully study the chapter, then answer the following questions.

__D__ 1. Broaching is a manufacturing process for machining _____.

A. flat, round, and contoured surfaces

B. internal and external surfaces

C. keyways, splines, and irregularly shaped openings

D. All of the above.

E. None of the above.

2. In broaching, a multitoothed cutting tool is _____ or _____ across the work. Each tooth on the broach removes a small amount of material.

pushed, pulled

3. Identify the parts of the broaching tool illustrated below.
   A. Finishing teeth
   B. Semi-finishing teeth
   C. Roughing teeth
   D. Pilot guide

A
B
C
D

4. List and briefly describe the three basic types of broaching operations.
   Internal broaching. Uses a pull broach.
   External broaching. Uses a slab broach.
   Pot broaching. The tool is stationary and the work is pushed or pulled through the broach.

D 5. Broaching is a very useful machining technique in that _____.
   A. consistently close tolerances can be maintained
   B. small parts can be stacked and shaped in a single pass
   C. it can remove metal faster than almost any other technique
   D. All of the above.
   E. None of the above.

D 6. A typical keyway broach set contains _____.
   A. instructions and a lubrication guide
   B. slotted bushings and necessary shims
   C. precision broaches
   D. All of the above.
   E. None of the above.

ram 7. A loose or worn arbor press _____ can damage a broach by pushing it to one side.

CHAPTER

18

The Milling Machine

Name ______________________ Date __________ Class __________

Learning Objectives

After studying this chapter, you will be able to:
- Describe how milling machines operate.
- Identify the various types of milling machines.
- Select the proper cutter for the job to be done.
- Calculate cutting speeds and feeds.

Carefully study the chapter, then answer the following questions.

___E___ 1. The milling machine is a very versatile machine tool. It can be used to _____.
A. drill, bore, and cut gears
B. machine irregularly shaped surfaces
C. machine flat surfaces
D. None of the above.
E. All of the above.

___cutter head___ 2. On a fixed-bed milling machine, vertical and cross movements are obtained by moving the _____.

___A___ 3. Rapid traverse feed _____.
A. permits work to be positioned at several times the fastest rate indicated on the feed chart
B. should only be activated while the cutter is positioned in a cut
C. allows fast power movement only when work is parallel with the periphery of the cutter
D. All of the above.
E. None of the above.

D 4. Climb milling should not be performed on machines _____.
A. without play in the table
B. that are not in top condition
C. not fitted with an antibacklash device
D. All of the above.
E. None of the above.

D 5. The ideal milling cutter should have _____.
A. edge toughness
B. red hardness
C. high abrasion resistance
D. All of the above.
E. None of the above.

D 6. High-speed steel milling cutters can be improved by the application of _____.
A. coatings such as chromium or tungsten
B. lubricating treatments
C. surface hardening treatments
D. All of the above.
E. None of the above.

B 7. Cemented tungsten carbides can be operated at speeds _____ than conventional HSS cutting tools.
A. much slower
B. 3–10 times faster
C. 10–14 times faster
D. up to 20 times faster
E. None of the above.

C 8. The term *hand* is used to describe _____.
A. cutter rotation
B. helix direction of the flutes
C. Both of the above.
D. Neither of the above.

9. _____ end mills can be fed into the work like a drill.
Two-flute

10. Multiflute end mill is recommended for conventional milling where _____ milling is not necessary.
plunge

11. Shell end mills have teeth similar to the multiflute end mills but are mounted on a(n) _____ arbor.
stub

Name ______________________________

12. What is a *fly cutter*?
A single-point cutting tool used as a face mill.

For questions 13–17 please match the milling cutters with their descriptions.

C 13. Mounts on an arbor and has cutting teeth only on its circumference.
E 14. Has helical teeth designed to cut with a shearing action.
B 15. Has cutting teeth on the circumference and on both sides.
D 16. Has alternate right-hand and left-hand helical teeth.
A 17. A thin milling cutter designed to machine narrow slots and for cutoff operations.

A. metal slitting saw cutter
B. side milling cutter
C. plain milling cutter
D. staggered-tooth side cutter
E. slab mill cutter

D 18. Formed milling cutters _____.
A. include corner rounding cutters
B. include concave and convex cutters
C. can be used to duplicate a contour
D. All of the above.
E. None of the above.

B 19. The Woodruff key seat cutter is used to mill _____.
A. the bottom of T-slots
B. semicircular keyseats
C. dovetail-type ways
D. All of the above.
E. None of the above.

shortest 20. In general, use the _____ arbor possible which permits adequate clearance between the arbor support and the work.

collars 21. Spacing _____ allow the cutter to be positioned on the arbor.

Drive keys 22. _____ fit into slots on the arbor, collet, or collet holder to provide nonslip drive.

23. Define the term *cutting speed*.
Cutting speed refers to the distance, measured in feet or meters, a point (tooth) on the cutter's circumference will travel in one minute.

24. Cutting speed is expressed in _____ or _____ per minute.
feet, meters

25. Define the term *feed*.
Feed is the rate at which the work moves into the cutter.

26. In what units is feed expressed?
Feed per tooth per revolution.

Using the formulas given below, solve problems 27–32. Use the space provided for your work. Round cutting speeds off to the nearest 50 rpm.

Formulas: $\text{rpm} = \dfrac{\text{fpm} \times 12}{\pi D}$

$F = \text{ftr} \times T \times \text{rpm}$

Where:
rpm = Revolutions per minute
D = Cutter diameter
ftr = Feed per tooth per revolution
T = Teeth in cutter

250 rpm 27. Calculate the machine speed (rpm) recommended for a 3″ diameter (8 teeth) side cutter (HSS) cutting free cutting steel. Recommended cutting speed is 200 fpm. Feed per tooth is 0.008″.

460 rpm 28. Determine the machine speed (rpm) for a 2.5″ diameter side cutter (HSS) with 8 teeth, milling brass. Recommended cutting speed is 300 fpm. Feed per tooth is 011″.

350 rpm 29. Calculate the machine speed (rpm) for a 6″ diameter side cutter (HSS) with 16 teeth, milling aluminum. Recommended cutting speed is 550 fpm.

Name ____

17800 rpm ____________ 30. Calculate the machine speed (rpm) for a 3/4″ diameter end mill (tungsten carbide) with 4 teeth, milling aluminum. Recommended cutting speed is 3500 fpm.

16 ipm ____________ 31. What will be the feed rate (F) for the job in Problem 30?

39.6 ipm ____________ 32. What will be the feed rate (F) for the job in Problem 31?

D ______ 33. Cutting speeds and feeds on a milling machine may be changed _____.

A. by shifting V-belts or by adjusting variable speed pulley
B. electronically
C. by using quick change gears
D. All of the above.
E. None of the above.

is not ____________ 34. When indexing, it _____ (is/is not) necessary to count holes each time the work is repositioned after a tooth has been cut.

D ______ 35. The rotary table can be used to cut _____.

A. irregular shaped objects
B. segments of circles
C. circular slots
D. All of the above.
E. None of the above.

dividing head ____________ 36. A(n) _____ can be used to measure the circumference of circular work into equally spaced units.

37. Identify the parts of the vertical milling machine illustrated below.

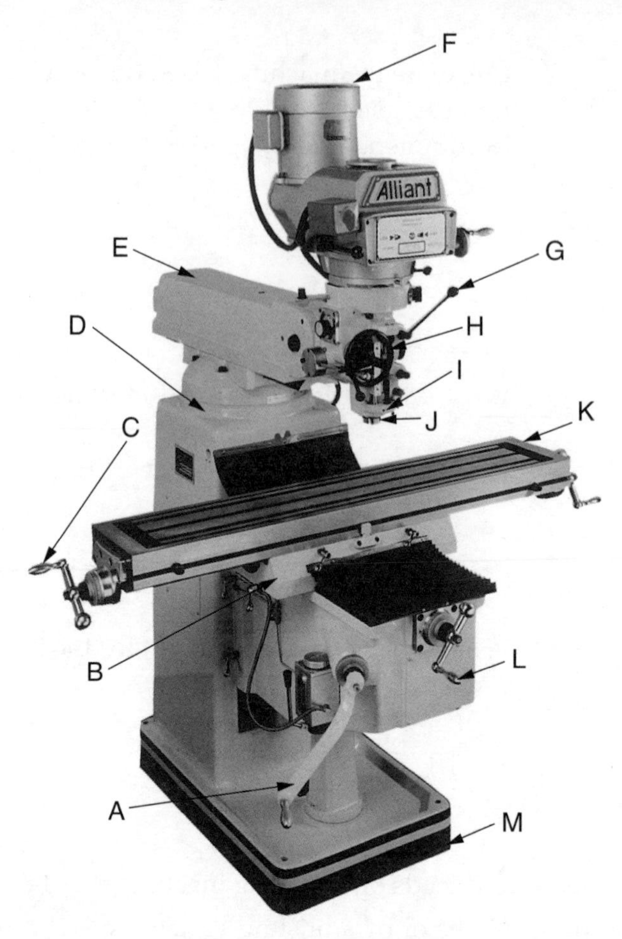

A. Vertical movement crank
B. Saddle
C. Longitudinal feed handwheel
D. Swivel
E. Overarm
F. Motor
G. Quill feed lever
H. Quill feed handwheel
I. Quill
J. Spindle
K. Worktable
L. Cross traverse handwheel
M. Base

38. List at least *four* safety precautions that must be observed when operating a milling machine.
Evaluate individually. Answers may include: Never operate machinery while your senses are impaired. Wear appropriate clothing and approved safety glasses. Stop the machine before making adjustments, measurements, or trying to remove accumulated chips. Be sure all power to the machine is turned off before opening or removing guards and covers. Use a piece of heavy cloth or gloves for protection when handling milling cutters. Get help when moving heavy machining equipment.

CHAPTER 19 Milling Machine Operations

Name ______________________ Date __________ Class __________

Learning Objectives

After studying this chapter, you will be able to:

- Describe how milling machines operate.
- Set up and safely operate horizontal and vertical milling machines.
- Perform various cutting, drilling, and boring operations on a milling machine.
- Make the needed calculations and cut spur gears.
- Make the needed calculations and cut a bevel gear.
- Point out safety precautions that must be observed when operating a milling machine.

Carefully read the chapter, then answer the questions in the space provided.

1. List four operations of which the vertical milling machine is capable of performing.
   Any order: milling, drilling, boring, reaming.
2. Why must the spindle of a vertical milling machine be at an exact right angle (perpendicular) to the worktable?
   If not, a flat surface cannot be machined.
3. What is the most accurate way to check for milling head perpendicularity?
   With a dial indicator and a special holder.
4. How should a vise be prepared for mounting to the worktable and holding work?
   By wiping the vise jaws and base clean and inspecting for burrs and ticks.

5. What can be used to check if the work is firmly seated in the vise?
Thin paper strips.

dial indicator 6. When extreme accuracy is required, the vise jaws must be aligned with a(n) _____.

D 7. Holes may be located to very close tolerances for drilling, reaming, or boring on a vertical milling machine with a(n) _____.
A. wiggler
B. edge finder
C. centering scope
D. All of the above.
E. None of the above.

against 8. When milling an internal slot that is wider than the cutter diameter, it is important that the direction of feed, in relation to cutter rotation, be observed. Feed direction is normally _____ cutter rotation.

9. Milling machine care is important to keep its accuracy. List five items that should be considered when preparing and operating the machine.
Evaluate individually. Refer to Section 18.4. Answers may include: clean the machine thoroughly after each job; check each setup for adequate clearance; never force a cutter into a collet or holder; always use a sharp cutter; make sure all guards are in place before attempting to operate the milling machine.

10. Why should the smallest diameter cutter that will do the job be used on a horizontal milling machine?
It is more efficient because it travels less distance while doing the same amount of work as a larger cutter.

toward 11. When using a helical slab mill, mount it so the cutting pressure forces it _____ (toward/away from) the column.

B 12. Face mills smaller than 6″ are _____ and are held on a Style C arbor.
A. held on a long arbor
B. called shell mills
C. called bell mills
D. All of the above.
E. None of the above.

Name ____

13. Explain *straddle milling*.
    A machining technique where a pair of cutters are used to machine both sides of the work at the same time.

14. How does *gang milling* differ from *straddle milling*?
    Gang milling employs mulitiple cutters to machine several surfaces in one pass.

climb 15. When slitting thin metal sections, best results can be obtained if the cutter is mounted for _____ milling.

16. Why is the above recommended?
    Forces the work down against the worktable.

17. How does *slotting* differ from *slitting*?
    In slotting, the cut is only made partway through the work.

are not 18. Inch-based gears and metric-based gears (are/are not) _____ interchangeable.

toward 19. When machining gears, cutting is done _____ (toward/away from) the dividing head.

eight 20. Gear cutters are made with _____ different forms for each diametral pitch.

21. When are bevel gears used?
    To change the angular direction of power between two shafts.

B 22. The tooth space at the pitch diameter of bevel gears is _____.
    A. the same at the small end as at the large end
    B. narrower at the small end than at the large end
    C. narrower at the large end than at the small end
    D. four times greater at the small end than at the large end
    E. None of the above.

D 23. Measurements of the finished bevel gear are made of the _____.
    A. depth of the tooth
    B. blank size and shape
    C. tooth thickness
    D. All of the above.
    E. None of the above.

24. How is the final inspection made after bevel gears are finished?
Final inspection is made by running the mating gears and checking for quietness and shape of the tooth contact.

Using the formulas from Section 18.9.2, find the information requested for problems 25–30. Perform your calculations in the space provided.

D_o = Outside diameter
D = Pitch diameter
a = addendum
t = Tooth thickness
h_t = Whole depth of tooth

25. Prepare the information to cut a 20 diametral pitch gear with 100 teeth. The dividing head has a 40:1 ratio and the index plate has the following series of holes: 33, 37, 41, 45, 49, 53, and 57.

D_o = 5.100″
D = 5.000″
h_t = 0.1078″
t = 0.0785″
a = 0.050″
Dividing head setup = 2/5 turn per tooth
Hole series used on index plate = 18 holes in 45 hole series in index plate
Gear cutter to be used = No. 2

Name ____

26. Prepare the information to cut a 4 diametral pitch gear with 18 teeth. The dividing head has a 5:1 ratio and the index plate has the following series of holes: 73, 77, 80, 85, 87, 90, and 95.

D_o = 5.000″

D = 4.500″

h_t = 0.539″

t = 0.392″

a = 0.250″

Dividing head setup = **5/18 turn per tooth**

Hole series used on index plate = **25 holes in 90 hole series in index plate**

Gear cutter to be used = No. 6

27. Prepare the information to cut an 8 diametral pitch gear with 12 teeth. The dividing head has a 5:1 ratio and the index plate has the following series of holes: 20, 23, 25, 29, 31, 33, 36, 39, and 41.

D_o = 1.750″

D = 1.500″

h_t = 0.2696″

t = 0.196″

a = 0.125″

Dividing head setup = **5/12 turn per tooth**

Hole series used on index plate = **15 holes in 36 hole series in index plate**

Gear cutter to be used = No. 8

28. Prepare the information to cut a 12 diametral pitch gear with 36 teeth. The dividing head has a 40:1 ratio and the index plate has the following series of holes: 60, 65, 67, 69, 72, 75, 81, and 85.

D_o = 3.1667″

D = 3.0000″

h_t = 0.1798″

t = 0.1309″

a = 0.0833″

Dividing head setup = 1 1/9 turns per tooth

Hole series used on index plate = 9 holes in 81 hole series in index plate

Gear cutter to be used = No. 4

29. Prepare the information to cut a 4 diametral pitch gear with 24 teeth. The dividing head has a 5:1 ratio and the index plate has the following series of holes: 96, 98, 100, 101, 104, 106, 109, 112, and 115.

D_o = 6.500″

D = 6.000″

h_t = 0.539″

t = 0.393″

a = 0.250″

Dividing head setup = 5/24 turn per tooth

Hole series used on index plate = 20 holes in 96 hole series in index plate

Gear cutter to be used = No. 5

Name ____________________

30. Prepare the information to cut a 3 diametral pitch gear with 56 teeth. The dividing head has a 40:1 ratio and the index plate has the following series of holes: 63, 65, 69, 73, 78, 83, 87, 90, and 95.

D_o = 19.333″

D = 18.667″

h_t = 0.719″

t = 0.524″

a = 0.333″

Dividing head setup = 5/7 turn per tooth

Hole series used on index plate = 45 holes in 63 hole series in index plate

Gear cutter to be used = No. 2

Notes

CHAPTER 20

Precision Grinding

Name ______________________ Date __________ Class __________

Learning Objectives

After studying this chapter, you will be able to:

- Explain how precision grinders operate.
- Identify the various types of precision grinding machines.
- Select, dress, and true grinding wheels.
- Safely operate a surface grinder using various work-holding devices.
- Solve common surface grinding problems.
- List safety rules related to precision grinding.
- Identify other types of precision grinding operations.

Carefully read the chapter, then answer the questions in the space provided.

1. What might a grinding wheel be compared to?
   A many-tooth milling cutter as each of the abrasive particles is a separate cutting edge.

hardness 2. Precision grinding is one of the few machining operations that can produce a smooth, accurate surface, regardless of the material's _____.

C 3. On a manually-operated surface grinder, the _____ handwheel controls up-and-down adjustment of the grinding wheel.

   A. traverse
   B. cross-feed
   C. downfeed
   D. All of the above.
   E. None of the above.

A 4. On a manually-operated surface grinder, the _____ handwheel controls the left-and-right movement of the table.

A. traverse
B. cross-feed
C. downfeed
D. All of the above.
E. None of the above.

B 5. On a manually-operated surface grinder, the _____ handwheel controls the in-and-out motion of the table.

A. traverse
B. cross-feed
C. downfeed
D. All of the above.
E. None of the above.

D 6. The control console of a surface grinder allows the operator to _____.

A. control dwell
B. start and stop table travel
C. adjust table speed
D. All of the above.
E. None of the above.

7. Much of the work done on a surface grinder is held in position by a(n) _____.
magnetic chuck

coolant 8. Wheel glazing or loading often indicates that the wrong _____ is being used.

9. Grinding wheels should be checked each time they are used. What is an easy way of doing this?
Tap it lightly with a metal rod. A solid wheel will give off a clear metallic ring.

10. What will occur if the grinding wheel is unbalanced?
Unbalanced wheels will cause irregularities on the finished ground surface.

11. Name at least three methods that can be used to apply cutting fluids.
Evaluate individually. Answers may include: By flooding the grinding area; using a mist system; manually applying with a pressure-type oil pump can.

Name ____________________________________

12. What problem will occur if the grinding wheel is out-of-round?
It can cause surface waviness.

13. Describe *creep grinding*.
Creep grinding is a surface grinding operation performed in a single pass with a depth of cut 1000–10,000 times deeper than conventional surface grinding.

14. What can cause irregular table movement or no table movement on hydraulic type machines?
Evaluate individually. Answers may include: clogged hydraulic lines; insufficient hydraulic fluid; hydraulic pump not functioning properly; inadequate table lubrication; cold hydraulic system; air in the system.

15. What can be the cause of work surface waviness? How can it be corrected?
The wheel being out-of-round. It can be corrected by truing the wheel.

D 16. Burning or work surface checking can be the result of _____.
A. wheel grain that is too fine
B. insufficient coolant reaching the work
C. a wheel that is too hard
D. All of the above.
E. None of the above.

A 17. Work that is not parallel is frequently caused by a(n) _____.
A. nicked or dirty chuck
B. wheel that is too coarse
C. chuck that has been *ground in* since the last time it was mounted
D. All of the above.
E. None of the above.

18. List at least four safety precautions that should be observed when doing precision grinding.
Evaluate individually. Answers may include: Never place a wheel on a grinder before checking it for soundness, check the wheel often to prevent it from becoming glazed or loaded, make sure the grinding wheel is clear of the work before starting the machine, never operate a grinding wheel at a speed higher than specified by the manufacturer.

Crowding 19. _____ the wheel into the cutter is a common mistake when grinding cutters because the lack of sparks creates the illusion that the cut being made is too light.

A 20. Irregular scratches on a ground surface are frequently caused by _____.

A. dirty coolant

B. an out-of-round grinding wheel

C. a glazed wheel

D. All of the above.

E. None of the above.

21. When sharpening cutters, what can cause more material to be removed from some teeth than others?

Evaluate individually. Answers may include any of the following: grinding wheel may be too soft and wearing down too rapidly; tooth rest may not be mounted solidly; the arbor may not be running true on the centers.

workhead 22. End mills are sharpened with the end mill mounted in a(n) _____.

twisting 23. Grinding cutters with helical teeth requires that a(n) _____ motion be used to keep the tooth correctly located against the grinding wheel.

radially 24. Form tooth cutters must be ground _____ to preserve tooth shape.

25. Describe *cylindrical grinding*.

Work is mounted between centers or in a chuck and rotates while in contact with the grinding wheel.

B 26. Table movement in the cylindrical grinding process should be adjusted so the wheel will overrun the work end by about _____ the width of the wheel face.

A. one-eighth

B. one-third

C. one-half

D. All of the above.

E. None of the above.

D 27. The internal grinding process _____.

A. is used to secure accuracy on inside diameters

B. uses a revolving wheel that moves in and out of the hole

C. is used to secure a fine surface finish

D. All of the above.

E. None of the above.

Name ________________________________

28. Describe *centerless grinding*.
    Evaluate individually. Answers may include the following: The work is rotated against the grinding wheel. It does not have to be supported between centers. The piece is positioned on a work support blade and fed automatically between a regulating or feed wheel and a grinding wheel. The regulating wheel causes the piece to rotate and the grinding wheel does the cutting. Feed through the wheels is obtained by setting the regulating wheel at a slight angle.

29. List the four variations of centerless grinding.
    Evaluate individually. Answers may include the following: Through feed, infeed, end feed, and internal centerless grinding.

30. In form grinding, the _____ is shaped to produce the required contour on the work.
    grinding wheel

31. List two advantages of abrasive belt machining.
    Evaluate individually. Answers may include any two of the following: Removes material at a high rate; run cool and require light contact pressure; versatility; belts may be used dry or with a coolant; reduce possibility of metal distortion caused by heat; soft contact wheels and flexible belt conform to irregular shapes.

E 32. The _____ on an abrasive belt grinder are usually made of cloth or rubber.
- A. serrated/slotted wheels
- B. platens
- C. contact wheels
- D. All of the above.
- E. Both A and C.

D 33. Applications of electrolytic grinding include _____.
- A. machining heat-sensitive work
- B. rapid removal of stock from alloy steel parts
- C. sharpening carbide tools
- D. All of the above.
- E. None of the above.

D 34. The function of _____ is programmed into a CNC grinder's computer.
A. vertical feed motion
B. linear feed rates and positioning
C. spindle start and stop
D. All of the above.
E. None of the above.

E 35. When using a CNC grinder, the operator must constantly _____.
A. monitor and control coordinate functions
B. adjust for grinding wheel wear
C. adjust the grinding wheel path
D. All of the above.
E. None of the above.

CHAPTER

21

Band Machining

Name ______________________________ Date ____________ Class ____________

Learning Objectives

After studying this chapter, you will be able to:

- Describe how a band machine operates.
- Explain the advantages of band machining.
- Select the proper blade for the job to be done.
- Weld a blade and mount it on a band machine.
- Safely operate a band machine.

Carefully read the chapter, then answer the following questions in the space provided.

1. Describe *band machining*.
   It uses a continuous saw blade for rapid chip removal. Each tooth is a precision cutting tool.

2. What safety precautions must be observed when handling band saw blades?
   Wear leather gloves and approved eye protection.

___C___ 3. Raker set is recommended for _____.
   A. work with varying thickness
   B. free cutting materials
   C. cutting large solids
   D. All of the above.
   E. None of the above.

gage 4. Extra strength can be obtained when using narrow blades by securing a blade of a heavier _____.

B 5. Contour cutting on a band machine should use _____.

A. the widest blade the machine will accommodate
B. the widest blade that will cut the desired curves
C. a flexible back blade
D. All of the above.
E. None of the above.

6. Why must several teeth be removed when preparing the band saw blade for welding?
To assure uniform spacing after the weld is made.

7. Why is it a good idea to develop a definite lubrication sequence when lubricating a band machine?
To reduce the possibility of missing a vital point.

B 8. _____ are recommended for continuous high-speed sawing.

A. Blade guide inserts
B. Roller guides
C. Narrow guides
D. All of the above.
E. None of the above.

crown 9. The center of the band should ride directly over the center of the wheel _____ on the rubber tire.

10. How is the necessary amount of blade tension determined?
By the width and pitch of the blade.

abrasive cloth 11. A band machine can be used for parts polishing by replacing the blade with a continuous band _____.

D 12. In friction sawing, _____.

A. the teeth scoop out the softened metal
B. only a small area is affected by the teeth
C. heavy pressure is used to cut ferrous metals
D. All of the above.
E. None of the above.

13. List two techniques that may be used to correct the problem of band teeth breaking out.
Evaluate individually. Answers may include any two of the following: Reduce feed pressure; use finer pitch band if thin material is being cut; be sure work is held solidly as it is being fed into band; use a heavier-duty cutting fluid.

Name ____________________

D 14. To increase cutting rate speed, you could _____.

A. use a coarser pitch blade
B. increase feed pressure
C. increase the band speed
D. All of the above.
E. None of the above.

15. List at least two corrections that can be made if the band does not run true against the saw guide backup bearing?

Evaluate individually. Answers may include any two of the following: Remove burr on back of band where joined; if hunting back and forth against backup bearing on guide, reweld blade with back of band in true alignment; check alignment of wheels; check backup bearing, replace if worn.

16. List at least two corrections that can be performed to prevent the premature dulling of band teeth.

Evaluate individually. Answers may include any two of the following: Use a slower cutting speed; replace the blade with a finer pitch band; be sure proper cutting fluid is used; increase feed pressure; check if band is installed with teeth pointing down.

17. List at least two corrections that can be done to prevent band teeth from breaking out?

Evaluate individually. Answers may include any of the following: Change to a heavier band; reduce cutting speed; check wheels for damage; if blade breaks at weld, use longer annealing time; reduce heat gradually; use finer pitch blade; reduce feed pressure.

low 18. If the band makes a belly-shaped cut, blade tension may be too _____.

guides 19. Cutting should not be started until the _____ have been set properly.

20. Identify the parts of the band saw blade illustrated below.

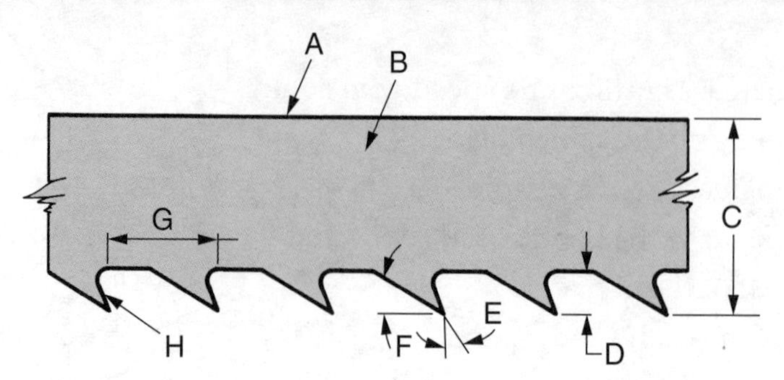

A. Back

B. Body

C. Width

D. Gullet depth

E. Tooth rake angle

F. Tooth clearance angle

G. Tooth spacing

H. Tooth face

CHAPTER

22

Introduction to CNC Machining

Name ______________________ Date __________ Class __________

Learning Objectives

After studying this chapter, you will be able to:

- Describe the development of CNC technology.
- List advantages and disadvantages of using CNC technology.
- Compare the characteristics of various types of CNC turning centers.
- Describe the features of CNC milling centers.
- Explain the relationship between CNC technology and flexible manufacturing systems.
- Identify safety guidelines for CNC machining processes.
- Summarize the use of the Cartesian and polar coordinate systems in CNC technology.
- Contrast the two types of motors that are commonly used to drive CNC machines.

Carefully read the chapter, then answer the following questions in the space provided.

1. Why were many companies slow to develop numerical control technology in the 1950s?
The cost of the machines and the skills needed to operate them outweighed the potential cost savings.

__C__ 2. The development of the minicomputer converted numerical control technology to ______.
   A. punched paper tape
   B. automated numerical control
   C. computer numerical control
   D. programmed codes
   E. None of the above.

3. Machining centers are equipped with a(n) _____ that automatically changes and stores the cutting tools.
automatic tool changer, ATC

C 4. A basic CNC milling machine has _____ axes of movement.
A. one
B. two
C. three
D. four
E. None of the above.

5. Explain how an operator uses conversational language to create programs for a CNC milling machine.
The operator uses software that comes with the control to select operations from menus.
The software converts the operator's input into standard programming code.

6. What are the advantages of a vertical machining center compared to a basic CNC milling machine?
Vertical machining centers typically have increased accuracy, precision, speed, and power.
In addition, they have automatic tool change capabilities and the ability to add rotational axes.

7. Designed to machine cylindrical parts, a(n) _____ is basically a computer-controlled lathe.
turning center

A 8. The least expensive type of turning center is the _____.
A. basic CNC lathe
B. gang-tool lathe
C. turret lathe
D. Swiss-type turning center
E. None of the above.

operator 9. For a basic CNC lathe, tooling changes are done by the _____.

10. Name two disadvantages of using basic CNC lathes.
CNC lathes are not suited for high-volume production. Their lack of precision and repeatability can create quality problems in high-precision manufacturing.

B 11. Which type of lathe has multiple tools mounted along the cross-slide?
A. Basic CNC lathe.
B. Gang-tool lathe.
C. Turret lathe.
D. Machining center.
E. None of the above.

Name ______________________________

___B___ 12. The cross-slide travel on a basic CNC lathe is usually limited to a distance of ______.

A. 6″ to 12″
B. 12″ to 18″
C. 18″ to 24″
D. 24″ to 30″
E. 30″ to 36″

13. List three advantages of gang-tool lathes compared to basic CNC lathes.
Evaluate individually. Answers may include: They can produce more parts per hour. The gang-tool setup allows for rapid tool changes. It also allows for low changeover times as long as the parts being machined are of similar diameter.

14. Explain how a turret lathe's cross-slide and turret work together during the machining process.
The turret rotates the appropriate tool into place and the cross-slide feeds the tool into the workpiece.

15. A Swiss-type turning center uses a(n) ______ to hold the workpiece tightly at the point of machining.
guide bushing

16. Why are the raw materials for Swiss-type turning centers more expensive than for other types of turning centers?
Because the guide bushing diameter is not adjustable, the stock used must be pre-ground to a precise diameter. This increases the cost of the raw materials significantly.

___automated___ 17. The only difference between CNC machines and manually operated machines is that CNC machines are ______.

18. Why is it important for an operator to know the sequence of operations executed by a CNC program?
Knowing the sequence allows the operator to recognize problems when the machine does something out of the ordinary.

___D___ 19. Which of the following auxiliary devices may be included on a CNC machine?

A. Coolant system.
B. Automated lubrication system.
C. Automated chip handler.
D. All of the above.
E. None of the above.

dry cycle 20. Cycling all programs during setup without a part in place is called a(n) _____.

C 21. What are the three axes of movement in the Cartesian coordinate system?

A. A, B, and C.
B. 1, 2, and 3.
C. X, Y, and Z.
D. I, II, and III.
E. R, θ, and Z.

22. Identify three methods used to transmit machining instructions to a CNC machine.

Evaluate individually. Answers may include: Instructions can be programmed directly through the keypad on the control, they can be uploaded using portable storage devices, or they can be uploaded through a computer network.

23. Describe one application where using the polar coordinate system is preferable to using the Cartesian coordinate system.

Evaluate individually. Answers may include: To locate holes in a circular pattern around a given point. Using the Cartesian coordinate system would require the programmer to use trigonometry and would involve rounding multiple decimal values.

B 24. The main drawback of the _____ is that it loses torque at higher speeds.

A. encoder
B. stepper motor
C. lead screw
D. servo motor
E. None of the above.

D 25. Which type of device requires an encoder to provide feedback about the position of the machine's axis to the control?

A. Lead nut.
B. Stepper motor.
C. Lead screw.
D. Servo motor.
E. None of the above.

lead screws 26. Motors drive the axes of a machine using _____.

backlash 27. Minimizing _____ is critical to all CNC machines to ensure precise positioning.

CHAPTER

23

CNC Programming Basics

Name ________________ Date ________ Class ________

Learning Objectives

After studying this chapter, you will be able to:

- Explain the process of planning and developing a CNC program.
- Describe three methods of generating CNC code.
- Summarize the use of CAD and CAM software in CNC programming.
- Classify the types of codes used in the ANSI/EIA 274D code format.
- Identify commonly used CNC modal commands.

Carefully read the chapter, then answer the following questions in the space provided.

1. Identify the positive and negative effects of the advances in CNC programming.
   Evaluate individually. Answers may include: Advances have enabled operators to create programs without needing in-depth programming knowledge. As a result, however, the ability of many operators to troubleshoot existing programs has declined.

print 2. In order to develop a CNC program, a programmer needs a copy of a part's _____.

C 3. In addition to the print, what other factor determines the necessary machining processes?
   A. Feed rate.
   B. Spindle speed.
   C. Raw material condition.
   D. Color-coding.
   E. None of the above.

4. How does color-coding a print help a programmer develop a CNC program?
Color-coding helps the programmer associate tooling and processes with the features on the drawing.

B 5. Which of the following is *not* a programming method used to generate CNC machine code?
   A. Manual data input (MDI).
   B. Paper tape programming.
   C. Conversational language programming.
   D. CAM programming.
   E. All of the above.

A 6. Which of the following programming methods involves typing code directly into a machine's control?
   A. Manual data input (MDI).
   B. Offline programming.
   C. Conversational language programming.
   D. CAM programming.

7. What are three advantages of conversational language programming compared to manual data input?
Evaluate individually. Answers may include any three: operators can enter information without the tedious practice of typing code, fewer mathematical calculations are needed, information can be entered directly from the print, multiple programs can be inputted more easily, in-depth knowledge of programming is not required.

depth 8. If a 2D CAD file is used as the basis for programming, the programmer must specify the _____ of the part.

9. What is *post-processing*?
The process of CAM software converting a CAM program into language that a CNC machine can understand.

10. The ANSI/EIA 274-D code format, commonly referred to as G-code, uses a _____ style of programming.
word address format

block 11. A single line of code is called a(n) _____.

Name ____

___A___ 12. The positioning of the machining cutter is controlled by _____.

A. G-codes
B. F-codes
C. M-codes
D. S-codes
E. E-codes

___C___ 13. Machine functions such as starting and stopping are controlled by _____.

A. G-codes
B. F-codes
C. M-codes
D. S-codes
E. E-codes

___D___ 14. Codes that have the same application no matter what control is being used are called _____.

A. temporary codes
B. permanent codes
C. unassigned codes
D. assigned codes
E. control codes

___D___ 15. In CNC codes, the number following the letter _____ determines the rpm of the spindle.

A. *G.*
B. *F.*
C. *M.*
D. *S.*
E. *E.*

16. What happens when the G00 command is activated?

When the rapid traverse command is called by the program to move, all axes will move at the fastest rate possible.

17. In CNC programming, straight-line movement is called _____.

linear interpolation

___**feed rate**___ 18. Every line of code that calls for a linear or circular interpolation move should include a _____ specification.

C 19. Which of the following commands signals circular movement in a counterclockwise direction?

A. G00.

B. G02.

C. G03.

D. G20.

E. G99.

20. How are I and J addresses used in combination with G02 or G03 command?

I and J addresses provide the location of an arc's center point. The numeric values after I and J specify the distance from the point where the machine is currently positioned to the center point of the arc.

modal 21. Once they have been activated, ______ commands remain activated until another command is activated.

B 22. Which of the following commands sets the control's units to millimeters?

A. G20.

B. G21.

C. G90.

D. G91.

E. G99.

23. What is the main disadvantage of incremental positioning if a program has to be modified?

Evaluate individually. Answers may include: If a point coordinate defined by incremental positioning must be changed, the coordinates of every point after the changed point must be changed as well.

CHAPTER 24

CNC Milling

Name ______________________ Date __________ Class __________

Learning Objectives

After studying this chapter, you will be able to:

- Identify the purpose of common miscellaneous function codes.
- Describe work-holding devices that are commonly used in CNC machining processes.
- List information the programmer needs to gather before beginning to write a CNC program.
- Carry out the procedure for writing a CNC milling program.

Carefully read the chapter, then answer the following questions in the space provided.

1. Besides allowing the operator to change tools, what other tasks can the M00 command be used for?
Evaluate individually. Answers may include: The M00 command can be used to stop a machine to allow for part or tool inspection or to clear chips away.

M01 2. If a CNC machine's toggle switch is set to the "off" position, the machine ignores the ______ command.

C 3. Which of the following commands signals that the end of a program has been reached, but does *not* reset the program.
   A. M00
   B. M01
   C. M02
   D. M05
   E. M30

4. What is the difference between the M03 code and the M04 code?
Both codes turn the spindle on, but the M03 code specifies clockwise rotation while the M04 code specifies counterclockwise rotation.

5. List five codes that can be used to stop the M03 code.
M00, M01, M02, M05, and M30.

D 6. Which letter is used in combination with the M06 command to signal the controller that the following number designates a turret position?
A. *G*.
B. *M*.
C. *S*.
D. *T*.
E. *V*.

7. What is the difference between the M07 code and the M08 code?
Both codes start the coolant flow, but the M07 code starts the coolant as a mist and the M08 code starts the coolant as a flood.

vise 8. A common work-holding device for CNC milling processes is the machining _____.

9. What is the main disadvantage of using a machining vise to hold parts?
It takes more time to remove finished parts and load un-machined parts for the next cycle.

twist 10. A spot drill is used to locate holes and to prevent a(n) _____ drill from walking when it comes in contact with the base metal.

11. In Cartesian coordinates, what is the location of the machine home?
X0.000, Y0.000, Z0.000

A 12. Which of the following codes is used to cancel the program zero function?
A. G53.
B. G54.
C. G55.
D. G56.
E. G99.

13. What is indicated by the following line of code: N0020 M06 T2?
The code indicates an automated tool change to the tool in the turret location designated as location 2.

Name ______________________________

semicolon 14. Besides parentheses, some machines allow the use of a(n) ______ to separate comment codes from actual program codes.

C 15. If a line of code reads *N0080 M03 S950*, which of the following statements is true?

A. The spindle will rotate counterclockwise.
B. The feed rate is 950 inches per minute.
C. The spindle will rotate at 950 rpm.
D. The program will reset automatically at the end of this line of code.
E. None of the above.

D 16. Which of the following codes initiates a canned cycle for a simple drilling operation?

A. G73.
B. G74.
C. G76.
D. G81.
E. G99.

R 17. In a line of code for a peck drilling canned cycle, the letter ______ is used to specify the position of the clearance plane to which the drill retracts.

18. In a line of code for a peck drilling canned cycle, the value for R is .200, the clearance value is .100, and the part thickness is .250. If the program uses five pecks, what is the value for Q? Show your work.
.110

lead 19. In a tapped hole, the distance from the crest of one thread to the crest of the next thread is called the ______.

Notes

CHAPTER 25

CNC Turning

Name ______________________ Date __________ Class __________

Learning Objectives

After studying this chapter, you will be able to:

- Identify work-holding devices that are commonly used with CNC lathes and turning centers.
- List the information needed to prepare for writing a program for a CNC turning center.
- Explain how to set the machine offsets and prepare the CNC machine for a turning operation.
- Carry out the procedure for writing a program to machine a part using a two-axis CNC lathe.

Carefully read the chapter, then answer the following questions in the space provided.

D 1. Which type of work-holding device is commonly used with CNC turning centers in a setting where parts are made in small quantities?

A. Three-jaw chuck.

B. Four-jaw chuck.

C. Collet.

D. Both A and B.

E. Both B and C.

collet 2. An automated system can be used to load and unload parts from a CNC turning center if the work-holding device is a(n) ______.

bar puller 3. A(n) ______ can be mounted to the turret and used to automate the turning process.

4. What is the first step in developing a turning program?

The first step of developing a turning program is to identify the processes to be performed and the necessary tools.

5. Why is it advisable to use different tools for the roughing and finishing operations in high-volume production operations?
   The roughing operation dulls the cutter faster than the finishing operation. Using different tools allows the roughing tool to be used longer, which may result in fewer cutters needed per production run.

A 6. On a CNC lathe, the *positive* direction of the Z axis runs from the _____ to the _____ end of the machine.
   A. headstock, tailstock
   B. tailstock, headstock
   C. cross-slide, spindle
   D. spindle, cross-slide
   E. None of the above.

.500 7. If an operation requires a 3.000″ diameter piece of stock to be reduced in diameter by 1.000″, the cutter must advance _____ inch(es) into the part.

8. Calculate the spindle speed required to machine a 2.000″ diameter piece of aluminum with a cutting speed of 100 feet per minute.
   200 rpm

B 9. Which of the following blocks of code sets the machine zero position?
   A. N0010 G00 G20 G90
   B. N0020 G54 X-3.50 Z4.50
   C. N0030 M05
   D. N0040 M06 T3
   E. None of the above.

facing 10. Typically, _____ is the first operation performed when turning a part.

11. CNC lathes have codes that perform operations similar to the canned cycles found on machining centers called _____.
   repetitive cycles

Name ______________________________

12. Explain the difference between the G71 code and the G 72 code.
    Both the G71 code and G72 code are used to remove large amounts of material, but the G71 code removes material by feeding the cutting tool along the Z axis, while the G72 code removes material by feeding the cutting tool along the X axis.

__C__ 13. Which of the following blocks of code would instruct the machine to use a repetitive cycle, feed the tool along the X axis during the cutting feed, remove .100" per pass, and retract .050" after each pass?

A. G71 U.100 R.050
B. G71 U.050 R.100
C. G72 U.100 R.050
D. G72 U.050 R.100
E. None of the above.

__finish profile__ 14. Before the second line of code for a G71 repetitive cycle can be completed, the ______ must be defined.

__offset__ 15. On the second line of code for a G71 repetitive cycle, the U and W letters define the ______, the amount of stock left for the finishing cycle.

16. What code is used after a G71 or G72 code to start the finish turning cycle?
    G70

__C__ 17. What code is used to perform a pecking cutoff or grooving cycle?

A. G73.
B. G74.
C. G75.
D. G76.
E. None of the above.

Notes

CHAPTER 26 Automated Manufacturing

Name ______________________ Date __________ Class __________

Learning Objectives

After studying this chapter, you will be able to:

- Define the term "automation."
- Summarize the traits of a flexible manufacturing system.
- Define the term "industrial robot."
- Identify the safety issues associated with automated manufacturing.
- Explain the similarities and differences between different rapid prototyping techniques.

Carefully read the chapter, then answer the following questions in the space provided.

__C__ 1. Automation is ______.

A. the use of a single computer to control all aspects of design, manufacturing, and distribution

B. a manufacturing system in which all machines are operated pneumatically

C. a method of production that employs a machine(s) to automatically perform one or more of the basic manufacturing processes

D. All of the above.

E. None of the above.

2. A grouping of machines used to perform multiple machining operations automatically is often called a(n) ______.

flexible manufacturing cell (FMC)

D 3. Work is transferred to and from the machines in a flexible manufacturing cell by ______.

A. robots or automated fixture carts
B. conveyors
C. specially designed loaders
D. All of the above.
E. None of the above.

smart 4. Flexible manufacturing systems make use of ______ tooling, which consists of cutting tools and work-holding devices that can be easily reconfigured to produce a variety of shapes and sizes within a given part family.

5. How does the JIT system work?
Parts and materials are scheduled for arrival at the time needed and not before.

6. What is one drawback of the JIT system?
Production can be reduced or stopped by weather delays or strikes.

7. An industrial robot consists of four basic components. Briefly describe each.
A. Controller: Performs computations for controlling the movement of the arm and wrist to the proper location.

B. Power supply: Supplies the power to operate the robot. May be hydraulic, pneumatic, or electric.

C. Manipulator: The articulated arm of the robot. The end of the arm is fitted with a wrist capable of rotational motion around one or more axes.

D. End-of-arm tooling: Device attached to the robot wrist for specific applications, such as a gripper, welding head, or spray gun.

Name ____________________

8. Briefly describe Laminated Object Manufacturing (LOM)™.
   Evaluate individually. Answers may include: CAD data is programmed into the LOM's process controller. A cross-sectional slice is generated and a laser cuts the outline of the cross section, then cross hatches the excess material for later removal. A new layer of material is bonded to the top of the previously cut layer, and the process continues until all layers are laminated and cut.

9. Briefly describe stereolithography.
   Evaluate individually. Answers may include: Data is downloaded into the stereolithography machine. The machine's control unit guides a fine laser beam onto the surface of a vat containing a liquid photocurable polymer, which solidifies wherever the laser beam strikes. After each slice is formed, the platform drops a small distance. The sequence is repeated until the entire model is formed.

10. Briefly describe Fused Deposition Modeling (FDM)™.
    Evaluate individually. Answers may include: A temperature-controlled head extrudes thermoplastic material layer by layer. The designed object emerges as a three-dimensional part without tooling. In a variation, parts made of sand or ceramic material, instead of plastic. Layers of special material are built up, using an inkjet-style printer to spray a quick-hardening binder to solidify each layer.

Notes

CHAPTER

27

Quality Control

Name ______________________ Date __________ Class __________

Learning Objectives

After studying this chapter, you will be able to:

- Explain the need for quality control.
- Summarize the difference between destructive and nondestructive testing.
- Describe the various nondestructive testing methods.

Carefully read the chapter, then answer the following questions in the space provided.

___prevent___ 1. The ultimate purpose of quality control is *not* to detect imperfect parts, but to ______ them from being made.

___D___ 2. Early advancements in quality control included the use of ______.

A. magnetic particle inspection
B. optical tools and jigs and fixtures
C. an increased number of inspectors
D. All of the above.
E. None of the above.

3. What is destructive testing?

A quality control process in which the part is destroyed during testing. The specimen is selected at random and does not ensure that untested parts are acceptable.

4. What is nondestructive testing?
A quality control process in which the usefulness of part is not impaired. Each piece is tested individually and as a part of a complete assembly.

5. Why must precision measuring tools be calibrated frequently?
Calibration ensures measuring tool accuracy by checking them against known standards.

6. A(n) _____ is an instrument that makes precise measurements electronically.
coordinate measuring machine (CMM)

7. How does an optical comparator work?
An enlarged image of the part being inspected is projected onto a screen where it is superimposed on a grid or accurate drawing overlay of the part. Very small size variations can be noted by skilled operator.

automates 8. An optical gaging system _____ the process of optical comparator inspection.

9. What is statistical process control?
A quality control program in which a percentage of all parts produced are measured. Statistical analysis is performed on the results of the inspections, which reveals developing problems before they result in unacceptable products.

10. What is radiographic inspection?
A nondestructive testing method that passes X rays and gamma radiation through a part and onto light sensitive film to detect flaws. The developed film has an image of the internal structure of the part.

D 11. Radiographic inspection offers which of the following advantages?
A. Parts can be inspected without taking them out of use.
B. Inspection sensitivity is high.
C. Internal and hidden parts can be inspected without damage.
D. All of the above.
E. None of the above.

Name ______________________________

12. Briefly describe the magnetic particle inspection process.
Evaluate individually. Answers may include: A magnetic field is induced in the part. Iron particles are blown or poured on the part. The iron particles outline flaws in the part.

13. What is a major disadvantage of magnetic particle inspection?
Evaluate individually. Answers may include: It cannot find flaws in nonferrous materials. It only detects serious defects.

14. How is a dye penetrant inspection performed?
A penetrant solution is applied to the surface of the part. Capillary action pulls the penetrant into the flaw. The surface is rinsed clean and a developer applied. The developer draws the dye to the surface, revealing flaws.

15. How is a visible dye penetrant inspection different from one conducted with a fluorescent dye?
Evaluate individually. Answers may include: The dye is visible under normal light rather than ultraviolet light, so the test does not require a black light. The developer provides a white background which sharply contrasts with the red dye.

sound waves 16. Ultrasonic inspection techniques use _____ to detect flaws in almost any kind of material.

D 17. Computer-controlled laser devices can be used to _____.
A. evaluate condition of products in use
B. detect tool wear
C. automatically compensate for tool wear
D. All of the above.
E. None of the above.

18. What is the basis of eddy-current inspection?
It is based on the fact that flaws in a metal product will cause impedance changes in a coil brought near it. Different eddy-currents will result in test coils placed next to metal parts with and without flaws. This difference determines which parts pass and fail inspection.

C 19. The eddy-current absolute system is used to detect _____.
A. flaws in metal parts
B. cracks, seams, or holes in metal parts
C. variations in dimension of a metal product
D. All of the above.
E. None of the above.

Notes

CHAPTER

28

Metal Characteristics

Name ______________________________ Date ____________ Class ____________

Learning Objectives

After studying this chapter, you will be able to:

- Explain how metals are classified.
- Describe the characteristics of different metals.
- Recognize the hazards that are posed when certain metals are machined.
- Explain the characteristics of some reinforced composite materials.

Carefully read the chapter, then answer the following questions in the space provided.

ferrous 1. Metals containing iron are classified as _____ metals.

2. What is a *base metal*?
A pure, nonferrous, nonprecious metal.

3. What is an *alloy*?
A mixture of two or more metals.

B 4. Carbon content in carbon steels is measured in percentage or points. Low carbon steel is defined as containing less than _____ percent or _____ points carbon.
A. 0.10, 10
B. 0.30, 30
C. 0.60, 60
D. 1.50, 150
E. 2.00, 200

D 5. Low-carbon steels _____.

A. cannot be hardened by conventional heat treatment
B. can be case hardened
C. can have a hard shell put on their surface
D. All of the above.
E. None of the above.

B 6. Cold finished steel is characterized by a _____.

A. black oxide coating
B. smooth, bright finish
C. flaky, yellow scale
D. All of the above.
E. None of the above.

C 7. High-carbon steels contain _____ percent or _____ points carbon.

A. 0.20 to .30, 20 to 30
B. 0.30 to .40, 30 to 40
C. 0.60 to 1.50, 60 to 150
D. 1.50 to 2.00, 150 to 200
E. 2.00 to 2.50, 200 to 250

D 8. High-carbon steels are _____.

A. used for a large variety of agricultural equipment
B. available in hot rolled form
C. used in products that must be heat-treated
D. All of the above.
E. None of the above.

9. Why are alloy steels more costly than carbon steels?
Evaluate individually. Answers may include: They are more costly to produce because of the increased number of special operations that must be performed in their manufacture.

C 10. _____ is added to steel when toughness, hardness and wear resistance are desired.

A. Nickel
B. Vanadium
C. Chromium
D. Tungsten
E. Molybdenum

Name ____

___B___ 11. _____ purifies steel and adds strength and toughness.

A. Molybdenum
B. Manganese
C. Chromium
D. Cobalt
E. Tungsten

___A___ 12. _____ imparts toughness and strength in steel, particularly at low temperatures.

A. Nickel
B. Vanadium
C. Cobalt
D. Tungsten
E. Molybdenum

13. What are *tool steels* and when are they used?

Evaluate individually. Answers may include: Tool steels found in devices that are used to cut, shear, or form materials. Tool steel may be either carbon or alloy steel. Steels in the lower carbon range are used for tools subject to shock. Steels in the higher carbon range are used when tools with keen cutting edges are required.

14. What is unique about tungsten carbide?

It is the hardest known metal.

___Magnesium___ 15. Which metal is classified as the lightest of the structural metals?

16. Stainless steels can be machined with techniques normal for mild steel. However, some precautions must be observed. What are three of these precautions?

Evaluate individually. Answers may include any three of the following: Feeds must be high enough to ensure that the cutting edge(s) get under the previous cuts; tools must be as large as possible; finishing cuts should be used when working to close tolerances; the machine should be adjusted so there is minimum play.

___D___ 17. _____ is a method used for identifying steels.

A. Color coding
B. The AISI code
C. Spark testing
D. All of the above.
E. None of the above.

nonferrous 18. Metals containing no iron are classified as _____ metals.

19. What are the two main classes of aluminum alloys?
Wrought and cast.

20. What does *temper designation* indicate when used in identifying the grade of aluminum?
It indicates the degree of hardness of an alloy.

21. Aluminum alloys possess many desirable qualities. List three of them.
Evaluate individually. Answers may include any three of the following: They are extremely strong and corrosion-resistant under most conditions; they are lighter than most commercially available metals; they can be shaped and formed easily; they are readily available in a multitude of sizes, shapes, alloys, and tempers.

22. What is a major safety problem when machining magnesium?
Evaluate individually. Answer may include: Magnesium alloys may contain thorium, a low-level radioactive material. Magnesium chips are highly flammable and burn at very high temperatures.

23. What type of coolants should be avoided when machining magnesium?
Water-based cutting fluids.

24. What precautions must be taken when machining beryllium copper?
A respirator-type face mask must be worn; special procedures must be followed when cleaning machines used to machine the alloy.

25. List two of titanium's many unique characteristics.
Evaluate individually. Answers may include any two of the following: Strong as steel, weighs only half as much as steel, extremely resistant to corrosion, most titanium alloys are capable of continuous use at temperatures up to about 800°F.

26. What is a *superalloy*?
Superalloy is another term for high-temperature metal. These are metals that maintain their high strength during extended periods at elevated temperatures.

Name ________________________________

___C___ 27. _____ melts at a higher temperature than any other known metal and is an ideal metal for breaker points in electrical devices.

A. Tantalum
B. Molybdenum
C. Tungsten
D. Titanium
E. Mercury

28. What are the properties of honeycomb sandwich panels?
Evaluate individually. Answers may include: They have a very high strength-to-weight ratio and rigidity-to-weight ratio.

29. Why are composites often used in place of many conventional metals?
Evaluate individually. Answers may include: They are generally lighter, stronger, and more rigid than many conventional metals. Some composites are capable of withstanding high temperatures.

30. How are composites made?
Fibers are bonded together in a plastic matrix under heat and pressure.

Notes

CHAPTER

29

Heat Treatment of Metals

Name ______________________ Date ____________ Class ____________

Learning Objectives

After studying this chapter, you will be able to:

- Explain why some metals are heat-treated.
- List some of the metals that can be heat-treated.
- Describe some types of heat treatment techniques and how they are performed.
- Explain how to case harden low-carbon steel.
- Summarize the process for hardening and tempering some carbon steels.
- Compare hardness testing techniques.
- List the safety precautions that must be observed when heat treating metals.

Carefully read the chapter, then answer the following questions in the space provided.

1. List the three changes obtained from heat treatment processes.
Improved resistance to shock, increased toughness, increased wear resistance and hardness

50 2. Steel with less than _____ points carbon cannot be hardened.

3. What purpose is the purpose of process annealing?
It removes internal stresses that have developed in parts that have been cold worked, machined, or welded.

4. What does the box annealing method involve and when is it used?
The part is placed in a metal box and the entire unit is heated. It is then allowed to cool slowly in the sealed furnace. The box annealing method is used to prevent the work from scaling or decarbonizing.

5. Briefly describe the purpose of annealing.
Annealing reduces the hardness of a metal, making it easier to machine or work.

6. Briefly describe the purpose of normalizing.
Normalizing refines the grain structure of some steels, thereby improving their machinability.

7. When is the heat treatment process of surface hardening most commonly used?
Surface hardening is used when a medium-hard surface is required on high-carbon or alloy steels.

8. How is the heat treatment process of hardening accomplished?
Hardening is accomplished by heating metal to its critical range and cooling it rapidly.

9. What is meant by the term critical temperature of steel?
Evaluate individually. Answers may include: It is the temperature at which steel's crystal structure changes; it is the temperature at which most heat treatments are performed.

B 10. _____ hardening involves the rapid heating of the surface with an acetylene torch.
A. Induction
B. Flame
C. Laser
D. All of the above.
E. None of the above.

11. List the three types of case hardening techniques.
Carburizing, cyaniding, nitriding

Name ________________________________

12. What is the effect of the case hardening process?
Case hardening puts a hard shell on the surface of low-carbon steel while the inner portion of the metal remains soft and tough.

carburizing 13. The ______ method of case hardening involves burying the steel in a material composed mostly of carbon.

14. What is the purpose of tempering.
Tempering reduces a metal's brittleness or hardness.

A 15. Some aluminum alloys must be refrigerated while they are worked in order to maintain their ______.
A. ductility
B. hardness
C. density
D. resistance to corrosion
E. conductivity

titanium 16. The heat treatment of ______ requires the use of special facilities to prevent the reactive metal from absorbing oxygen, carbon, and nitrogen.

D 17. Which of the following is a common quenching medium?
A. Brine.
B. Cold air.
C. Oil.
D. All of the above.
E. None of the above.

18. Why are some furnaces equipped with a second, high-temperature chamber?
The second chamber can be used to provide atmospheric control for heat treating with inert gases.

19. How can the temperature of the metal be determined if a furnace is *not* equipped with a pyrometer?
As the metal is heated, its color can be compared to a color chart to determine its temperature.

carburizing 20. The ______ method of case hardening uses a nonpoisonous case hardening compound.

standards ______ 21. Hardness testing establishes ______ for hardness that can be cited on drawings and specifications.

22. Why are the Brinell and Rockwell hardness testing machines called indention hardness testers?
They are used in techniques that measure the distance a steel ball or special-shaped diamond penetrates into the metal under a specific load.

23. How does the scleroscope hardness tester differ from the Brinell and Rockwell testers?
The scleroscope technique determines hardness based on the rebound height of a hammer dropped on the specimen.

A ______ 24. The ______ is a portable testing device that can be used on assemblies that cannot be brought into the laboratory or to test a variety of shapes, such as extrusions, tubing, or flat stock.

A. Webster hardness tester
B. Rockwell hardness tester
C. Brinell hardness tester
D. All of the above.
E. None of the above.

cyanide ______ 25. Potassium ______ should never be used in a school shop or lab as a case hardening medium because the fumes it produces are extremely toxic.

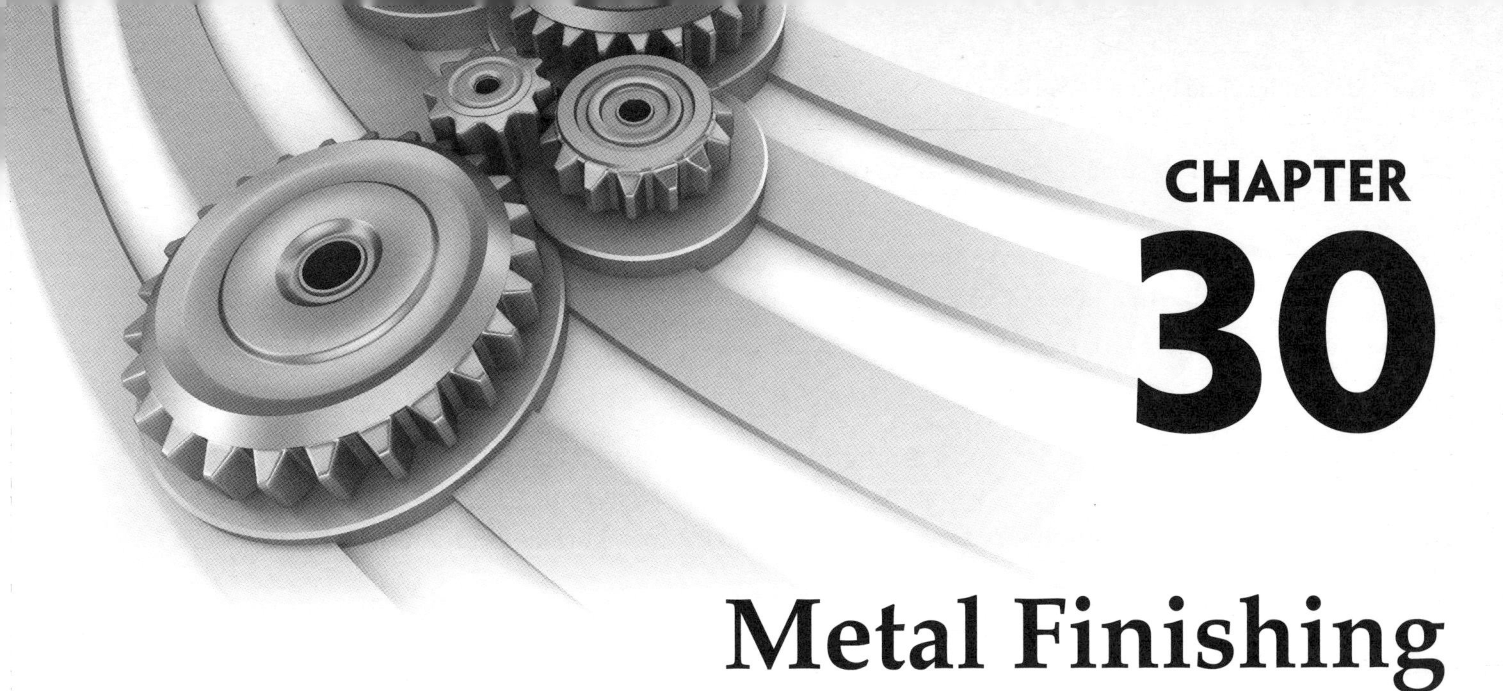

CHAPTER

30

Metal Finishing

Name ______________________ Date __________ Class __________

Learning Objectives

After studying this chapter, you will be able to:

- Describe how the quality of a machined surface is determined.
- Explain why the quality of a machined surface has a direct bearing on production costs.
- Identify organic coatings and describe how they are applied.
- List common inorganic coatings and summarize how they are applied.
- Describe different methods of applying metal coatings.
- Explain various methods of deburring.

Carefully read the chapter, then answer the following questions in the space provided.

1. Before standards were established to define the different degrees of surface roughness, drawings included this information in notes. What were the problems caused by this way of specifying machined surface finishes?
   Each machinist interpreted specifications differently and pieces were often better finished than necessary, raising costs.

D 2. The terms and ratings of surface roughness standards, or texture standards, relate to surfaces produced by _____.
   A. machining
   B. grinding or casting
   C. molding or forging
   D. All of the above.
   E. None of the above.

D 3. The surface roughness standards deal only with the _____ of surface irregularities.

A. direction

B. width

C. height

D. All of the above.

E. None of the above.

irregularities 4. Surface roughness standards arrive at roughness values by averaging, arithmetically, the size of _____ on the surface.

5. Surface roughness can be checked visually with a microfinish surface comparator, also called a(n) _____.

surface roughness gage

6. Surface roughness can be checked electronically with a(n) _____, which is more accurate than visual methods.

profilometer

7. Waviness is another surface condition. What does it describe?

The smoothly rounded peaks and valleys caused by tool vibration and chatter.

B 8. Lay is another surface condition that can be measured. It describes the _____.

A. peaks and valleys caused by machine vibration and chatter

B. direction of predominate tool marks, grain, or pattern of surface roughness

C. appearance of the machined surface

D. All of the above.

E. None of the above.

higher 9. The finer the quality of a surface finish specified, the _____ the cost of producing it.

10. In addition to the surface finish of machined surfaces, other finishing techniques are also used by the metalworking industry for one or more different reasons. List and briefly describe two of them.

Evaluate individually. Answers may include: To enhance the appearance, to increase corrosion or wear resistance, to make product identifiable, to reduce production cost.

Lapping 11. _____ is used by automotive manufacturers to produce mating surfaces because it produces the smoothest finish on a production basis.

Name ______________________________

12. List four finishes included in the organic coating category.
    Evaluate individually. Answer may include the following: paints, varnishes, lacquers, enamels, various plastic-base materials, epoxies.

13. What is an anodized coating?
    A protective layer of aluminum oxide formed on aluminum parts.

14. What are the three classes of anodizing?
    Ordinary anodizing, hard coat anodizing, electrobrightening

15. What type of anodizing occurs when the electrolyte dissolves the oxide film at about the same rate that it is formed?
    Electrobrightening

Match the finishing techniques on the right to the statement on the left that best describes it. Place the letter of the technique in the appropriate blank.

Answer	Statement	Technique
F	16. A power polishing operation.	A. Anodizing.
G	17. Used to remove burrs.	B. Vitreous enamel.
A	18. A process that forms a protective layer of oxide on aluminum parts.	C. Chemical blackening.
B	19. A glass coating fused to sheet or cast iron surfaces.	D. Electroplating.
C	20. Often used to reduce light glare from parts and tools.	E. Metal spraying.
D	21. Metal coating deposited on metal surfaces by an electric current.	F. Buffing.
E	22. Metal wire or powder heated to its melting point and blown onto work surface.	G. Power brushing.

23. Briefly describe how porcelain is fused to sheet or cast iron surfaces.
    It is applied as a powder (frit), or as a thin slurry (slip). After the finish dries, the material is fired at about 1500°F (815°C) until it fuses to the metal surface.

24. Why is chemical blackening done?
    Evaluate individually. Answers may include: Enhances appearance of part; protects machined surface against humidity and corrosion; reduces glare; abrasive resistance is improved; adhesion qualities are improved.

25. What are two common reasons for applying flame-sprayed coatings?
To build up worn or scored surfaces so they can be remachined to required size, and to apply superhard coatings where abrasion-resistant surfaces are needed.

26. What is the detonation gun coating process?
Evaluate individually. Answers may include: It is a process for depositing metallic coatings on a workpiece. It uses a water-cooled barrel several feet long and about one inch in diameter that is fitted with valves for introducing gases and the coating material. It uses gases detonated in the barrel to melt the coating materials and propel them toward the part to be coated.

27. What are some advantages of the detonation gun coating process?
Evaluate individually. Answers may include: It can be fully automated; it can be used to apply coatings with high melting points to fully heat-treated parts without danger of changing the metallurgical properties or strength of the part and without danger of thermal distortion; and almost any material that can be melted without decomposing can be sprayed.

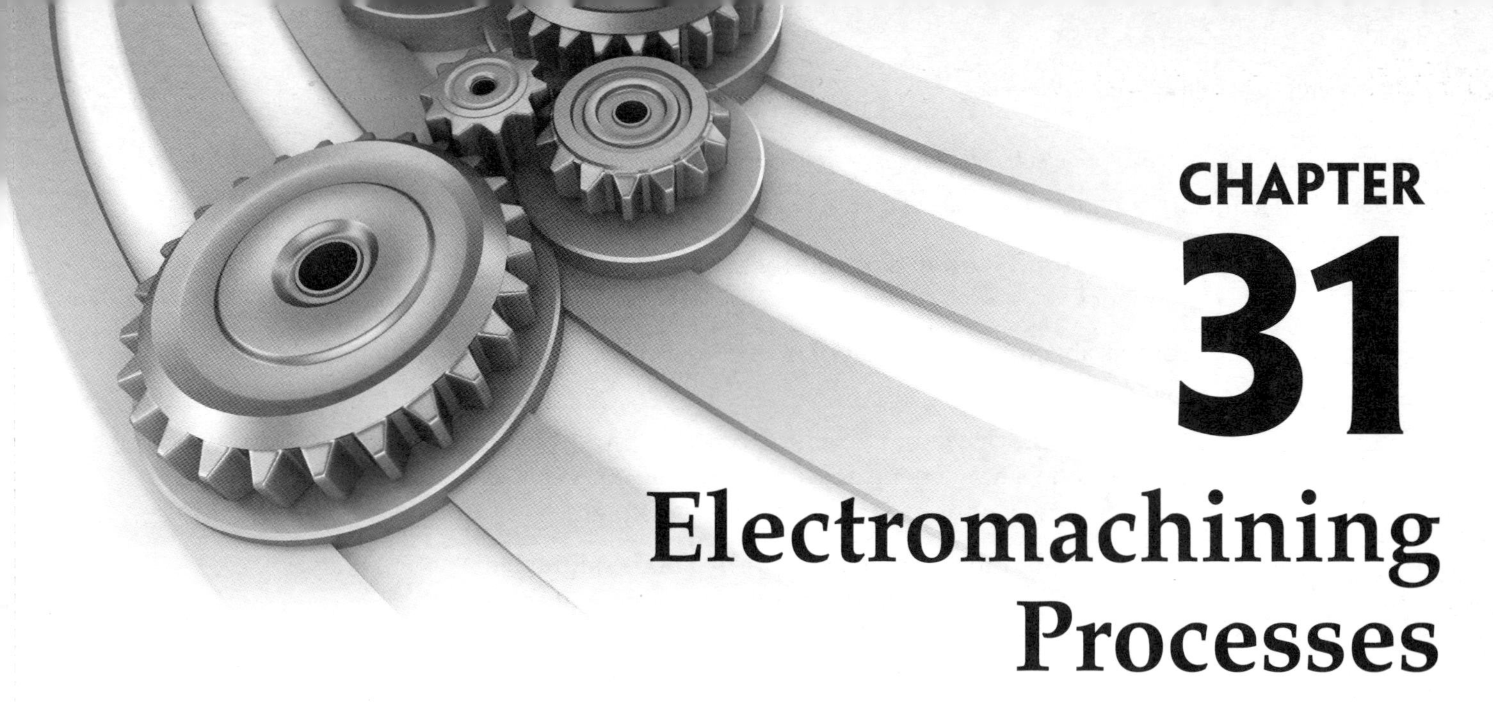

CHAPTER 31

Electromachining Processes

Name ______________________ Date __________ Class __________

Learning Objectives

After studying this chapter, you will be able to:

- Explain the advantages and disadvantages of the electromachining processes.
- Describe electrical discharge machining.
- Explain electrical discharge wire cutting.
- Summarize the small hole EDM drilling process.
- Describe electrochemical machining.

Carefully read the chapter, then answer the following questions in the space provided.

___C___ 1. A notable characteristic of electromachining is that _____.
   A. it can be used to work any material
   B. heat from the process automatically hardens the workpiece
   C. metal removal is by erosion and there are no chips
   D. All of the above.
   E. None of the above.

___conduct electricity___ 2. In electromachining, the metal must _____ in order to be machined.

___D___ 3. EDM is a process that permits the machining of metals that are too _____.
   A. difficult to work to close tolerances by conventional machining
   B. tough to be machined by conventional methods
   C. fragile or heat sensitive
   D. All of the above.
   E. None of the above.

Match the four main parts of an electrical discharge machine on the right to the statement on the left that best describes it. Place the letter of the part in the appropriate blank.

A. Dielectric fluid.
B. Power supply.
C. Servomechanism.
D. Electrode.

B ______ 4. Provides direct current and controls voltage and frequency.

D ______ 5. Comparable to the cutting tool on a conventional machine tool.

C ______ 6. Also called the drive unit, used to accurately control the electrode's movements and maintain the correct distance between the electrode and the workpiece.

A ______ 7. Usually a light mineral oil used to form a non-conductive barrier between the electrode and the workpiece at the arc gap.

B ______ 8. In the electrical discharge machining process, ______.
- A. the electrode is in direct physical contact with the workpiece
- B. no chips are produced as metal is removed
- C. the electrode and workpiece are submerged in highly conductive electrolyte
- D. All of the above.
- E. None of the above.

A ______ 9. ______ is the most used material for EDM electrodes.
- A. Graphite
- B. Brass
- C. Tool steel
- D. Copper
- E. Aluminum

A ______ 10. In the electrical discharge machining process, ______.
- A. roughing cuts are made at low voltage and low frequency
- B. finishing requires high amperage with high capacitance
- C. roughing cuts are made at high voltage and low frequency
- D. finishing requires low amperage with high capacitance
- E. All of the above.

slower ______ 11. Hard metals erode at a ______ (faster/slower) rate than soft metals.

D ______ 12. The EDM process is used to shape carbide tools and dies and to ______.
- A. machine complex shapes in hard, tough metals
- B. eliminate tedious and expensive hand work
- C. drill holes that are other than round in shape
- D. All of the above.
- E. None of the above.

Name ______________________________

13. How does EDWC differ from EDM?
A small-diameter wire is used as the electrode.

gang-cut 14. The EDWC process allows layers of sheet metal to be _____ to produce a number of parts in one pass.

C 15. In small hole EDM drilling, a high-pressure _____ is pumped through the center of the hollow electrode.
   A. electrolyte
   B. abrasive slurry
   C. dielectric fluid
   D. acidic solution
   E. None of the above.

D 16. Which of the following statements regarding small hole EDM drilling is *not* true?
   A. The electrode never comes into direct physical contact with the workpiece.
   B. A new or freshly dressed electrode must be used at the bottom of a blind hole.
   C. The hollow electrode spins as it progresses through the workpiece.
   D. Deep holes created by small hole EDM drilling tend to drift more than those created by mechanical drilling.
   E. All of the above.

C 17. ECM is an electromachining technique in which _____.
   A. metal is removed slowly
   B. metal hardness greatly affects speed of metal removal
   C. there is no tool wear
   D. All of the above.
   E. None of the above.

salt 18. The electrolyte for ECM is usually common _____ mixed with water.

19. What are the differences between ECM and EDM?
Evaluate individually. Answers may include: The fluid between the electrode and workpiece is an electrolyte rather than a dielectric fluid; there is no spark created between the workpiece and electrode; material is removed by electrolysis rather than arc erosion.

A 20. ECM removes metal rapidly, up to _____ cubic inch(es) per minute for each 10,000 amps of machining current.

A. one
B. two
C. three
D. four
E. five

21. What are three advantages of ECM?

Evaluate individually. Answers may include any three of the following: Metal is removed rapidly; the kind of metal or its hardness does not affect the speed of material removal; it is accurate; the machined metal is stress-free; there is no tool wear; it can eliminate the need for several other machining operations; advances are continually being made in ECM technology.

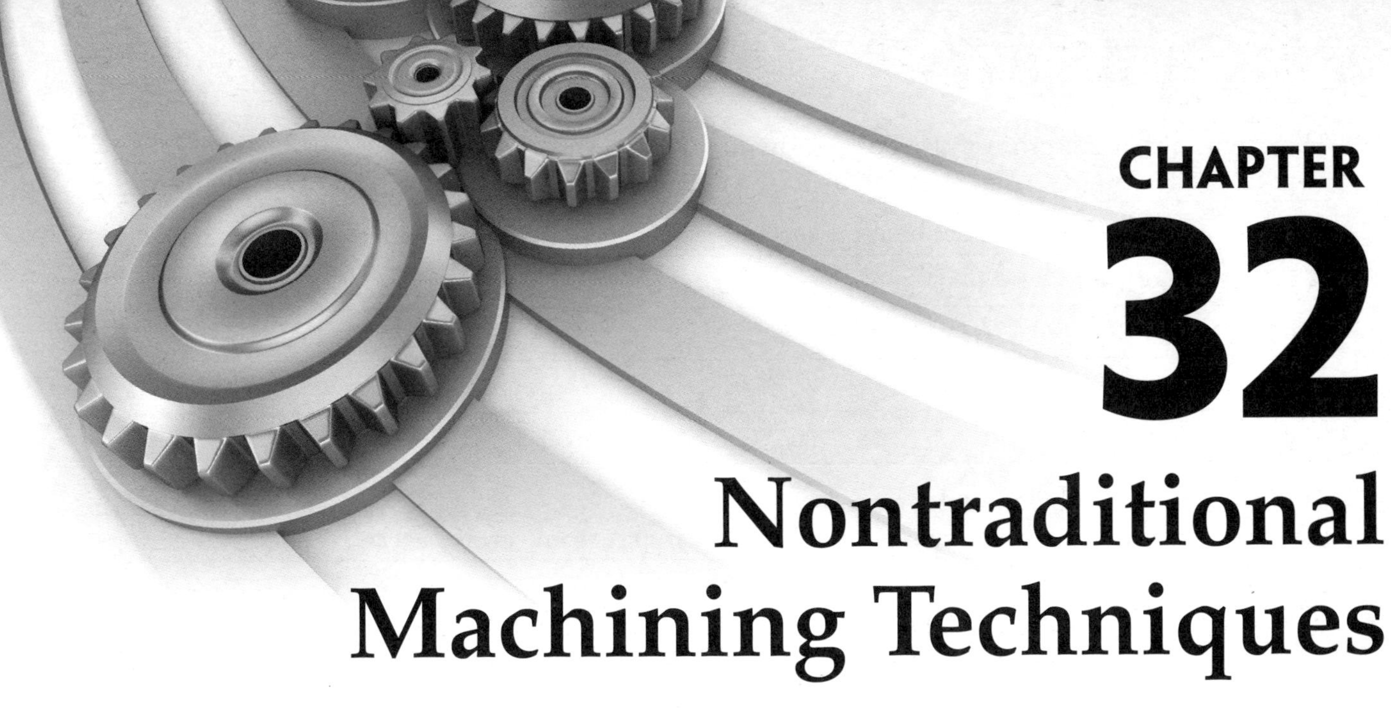

CHAPTER

32

Nontraditional Machining Techniques

Name ______________________ Date __________ Class __________

Learning Objectives

After studying this chapter, you will be able to:

- Describe the chemical milling process and its advantages and disadvantages.
- Explain water jet cutting and water jet abrasive milling.
- Summarize the various ultrasonic machining processes.
- Explain electron beam machining.
- Describe the laser beam machining process.

Carefully read the chapter, then answer the following questions in the space provided.

1. What is the principle of chemical machining?
   Chemicals, usually in an aqueous (with water) solution, are used to etch away selected portions of the metal to produce an accurately contoured part.

2. In chemical machining, the area not subject to metal removal must be protected by a ______.
   mask

3. List three of the many advantages that chemical milling offers.
   Evaluate individually. Answers may include any three of the following: Tooling costs are low; tight tolerances are possible; the size of the workpiece size is limited only by the size of the immersion tank; warping and distortion of formed sections is negligible; contoured or shaped parts can be chem-milled after they are formed; many parts can be produced simultaneously; thin metal can be machined without buckling; both sides of the metal can be milled at the same time; heat-treated metals can be machined chemically; no burrs are produced in the machined area.

4. The chemical milling process also has disadvantages. List three of them.
Evaluate individually. Answers may include any three of the following: The process is slow; all surface imperfections must be removed before etching; it is not recommended for etching holes; surface finishes on deep etches are not as fine as conventionally machined surfaces; lateral dimensions are difficult to hold.

5. Briefly describe the chem-blanking process.
Chemical blanking involves complete removal of metal from certain areas by chemical action. It is a variation of chemical milling.

jet 6. Hydrodynamic machining (HDM) is the technical name for water _____ cutting.

7. Why was hydrodynamic machining first developed?
To shape composites.

hard 8. Abrasives are added to the water jet for cutting metal and other _____ materials.

9. During water jet abrasive milling operations, how are contours created in the workpiece?
A mask is applied to areas where no metal removal is desired; the depth of the cut is determined by the speed at which the workpiece is fed.

D 10. The ultrasonic machining process uses sound waves _____.
A. by applying them directly to the cutting tool
B. in conjunction with a fluid
C. by applying them directly to the metal as it is being machined
D. All of the above.
E. None of the above.

11. What is ultrasonic-assist machining?
Ultrasonic-assist machining applies sound waves to the tool or metal as it is cutting or being cut; the process reduces tool forces and almost completely eliminates tool chatter.

Name ____

__C__ 12. Impact machining is done by a shaped cutting tool oscillating about _____ times per second.

A. 5,000
B. 15,000
C. 25,000
D. 45,000
E. None of the above.

13. What is the purpose of the funnel-shaped horn used in the impact machining process?
It amplifies and transmits ultrasonic waves from the transducer.

__high vacuum__ 14. EBM can cut any known metal or nonmetal that can exist in a(n) _____.

__magnetic lens__ 15. In EBM, the electron beam is tightly focused by a(n) _____.

__A__ 16. The pulsing technique used in the EBM process, requires the beam to be _____.

A. off longer than it is on
B. on longer than it is off
C. on and off for equal amounts of time
D. on continuously, but varying in intensity
E. None of the above.

17. In EBM, how is cut geometry controlled?
Cut geometry is controlled by movement of the worktable in the vacuum chamber and by employing the deflection coil to bend the beam of electrons to the desired cutting path.

__microns__ 18. Lasers produce a narrow beam of intense light that can be focused on an area only a few _____ in diameter.

19. Temperatures up to _____ can be instantaneously created at the point of focus of a laser beam.
75,000°F (41 650°C)

__B__ 20. Using a laser in the pulse mode _____.

A. provides a better edge quality
B. concentrates heat in localized area
C. keeps the cut on the programmed path
D. All of the above.
E. None of the above.

Notes

CHAPTER 33

Other Processes

Name ______________________________ Date ____________ Class ____________

Learning Objectives

After studying this chapter, you will be able to:

- Discuss the general machining characteristics of various plastics.
- Describe the hazards associated with machining plastics.
- Describe the five basic operations of chipless machining and their variations.
- Explain how the Intraform process differs from other chipless machining techniques.
- Describe how powder metallurgy parts are produced.
- Compare the advantages and disadvantages of various HERF techniques.
- Explain how the science of cryogenics is used in industry, and list some applications.

Carefully read the chapter, then answer the following questions in the space provided.

1. Why must special care be taken when machining many plastics?
Dust and fumes given off by some plastics may be irritating to the skin, eyes, and respiratory system. Other plastics have fillers, such as glass fibers, that are harmful to your health.

2. List four of the physical characteristics of nylon?
Evaluate individually. Answers may include any four of the following: High tensile strength, impact resistant, flexural strength, resistance to abrasion, not affected by most chemicals, greases, and solvents.

supported ______ 3. Since nylon is not as rigid as metal, it must be _____ during machining.

soft brass ______ 4. Most types of nylon can be machined using the same techniques used to machine _____.

5. Why must cutting tools be kept sharp when machining plastics?
Tools must be kept sharp to prevent the plastic from melting or becoming gummy. Sharp tools also ensure a good quality surface finish.

annealing ______ 6. Like some metals, machined nylon parts require _____ to prevent dimensional changes.

7. What are the desirable properties of acetal resin and what are some of its applications?
It has excellent dimensional stability, high strength, and rigidity. It has low friction, requires minimal use of lubricants, and is very quiet in operation. It is replacing brass and zinc for many applications in the automotive and plumbing industries and is used for parts in business machines.

A ______ 8. Teflon® has a high thermal expansion rate. When tolerances are critical, measurements should be made when the plastic is at _____.

A. the temperature at which it will be used
B. 10° below room temperature
C. a temperature of 74°F (21°C)
D. All of the above.
E. None of the above.

9. When Teflon® is being machined, what problem can be caused by a buildup of chips around the work?
A buildup of chips can prevent the heat from dissipating.

10. Why are large amounts of water-based coolant required during the machining of Teflon®?
To deter thermal expansion.

11. What is *chipless machining*?
Chipless machining forms wire or rod into the desired shape using a series of dies.

Name ______________________________

12. While chipless machining will not replace conventional machining, it offers several advantages for some types of jobs. What are the advantages?
Cost saving on some jobs, scrap is reduced, and increased production speed.

spark plug 13. Almost all _____ bases are made by chipless machining.

14. What is *sintering*?
Sintering is another name for powder metallurgy, a technique used to shape parts from metal powders. It is also the process of transforming the briquette into a strong unit.

15. Why must the briquette or green compact be handled carefully?
A briquette is brittle and very fragile.

C 16. The term *spring back* means _____.
A. using HERF to make metal springy
B. shaping metal against the back edge of a die
C. the metal tries to return to its original shape
D. All of the above.
E. None of the above.

17. Briefly describe stand-off HERF operations.
In stand-off HERF operations the charge is located some distance above the work. Its energy is transmitted through a fluid medium, such as water.

18. Briefly describe contact HERF operations.
With contact HERF operations the charge is touching the work and the explosive energy acts directly on the metal.

19. How is *magnetic forming* accomplished?
An insulated coil is wrapped around or placed within the work. As very high momentary current is passed through the coil, an immense magnetic field is created, causing the work to collapse, compress, shrink, or expand depending on the design of the coil.

20. Briefly describe *pneumatic-mechanical forming*.

High-pressure gas is used to accelerate a punch into a die. The forces developed are many times more powerful than those used in conventional forging and are sufficient to shape hard-to-work materials. The metal blank is heated prior to the forming operation and the machine requires less space than the conventional forging press.

B 21. Cryogenic treatment of cutting tools involves _____.

A. the immersion of cutting tools in liquid nitrogen
B. the use of liquid nitrogen as a gas
C. cooling the tools very rapidly
D. All of the above.
E. None of the above.